Yoke Leen Sit

MIMO OFDM Radar-Communication System with Mutual Interference Cancellation

Karlsruher Forschungsberichte
aus dem Institut für Hochfrequenztechnik und Elektronik

Herausgeber: Prof. Dr.-Ing. Thomas Zwick

Band 82

MIMO OFDM Radar-Communication System with Mutual Interference Cancellation

by

Yoke Leen Sit

KIT Scientific Publishing

Dissertation, Karlsruher Institut für Technologie (KIT)
Fakultät für Elektrotechnik und Informationstechnik, 2016

Impressum

KIT Scientific Publishing

Karlsruher Institut für Technologie (KIT)
KIT Scientific Publishing
Straße am Forum 2
D-76131 Karlsruhe

KIT Scientific Publishing is a registered trademark of Karlsruhe Institute of Technology. Reprint using the book cover is not allowed.

www.ksp.kit.edu

Print on Demand 2017 – Gedruckt auf FSC-zertifiziertem Papier

ISSN 1868-4696
ISBN 978-3-7315-0599-0
DOI 10.5445/KSP/1000061697

Vorwort des Herausgebers

At the turn on the new millennium, the radar technology underwent a radical change. What previously has been employed exclusively for the long range detection of objects (e.g. airspace surveillance), within a short span of 15 to 20 years later, evolved to include the applications for shorter range detection of one meter to a few hundred meters, which also increasingly became more economically relevant. With their high production volume and high complexity, the introduction of radar sensors into the automotive sector in particular has induced a clear change in radar technology. Single chip frontends are now available, where the complete RF circuitry is realized on an SiGe chip. Interconnect technologies meanwhile have been highly optimized to the standard of SMD assembly at 76 GHz. These hardware developments mark the beginning of the radar revolution. Different studies estimate that road vehicles equipped with at least one radar will hit 30% to 80% in year 2030. Due to the continuous increase in demand, a development similar to that experienced by the communication technology in the past 25 years is envisioned. It is expected that after the integration of the frontends, highly integrated signal processing ASICs will become available which allow a dramatic increase in complexity of the algorithms without increasing the cost. Just as with the transition of the mobile communication technology from the second to the third and fourth generations, more complex and sophisticated modulation techniques and algorithms will be used. This is where the work of M.Sc. Yoke Leen Sit touches on.

In her thesis, Ms. Sit provided important scientific fundamentals to the OFDM-based MIMO radar system with simultaneous communication capability with the focus on interference suppression techniques. Ms. Sit developed a complete signal model for multiple simultaneous system nodes in operation, each comprising multiple transmit and receive antennas. Hence a subcarrier allocation strategy for a networked scenario is also included. Based on the software defined radio modules (USRPs), Ms. Sit has set up a demonstrator, where for the first time, verification of configurations become possible. The major contribution of this work is the novel algorithm for interference cancellation. Contrary to communication systems, where the correct reception and decoding of signals from the multiple transmitters are of importance, interference cancellation for a radar demands a much higher accuracy in the estimation of all channel coefficients. Through the smart modification of well-known algorithms, Ms. Sit succeeded in improving the state of the art. With the demonstrator, Ms. Sit has also managed to verify all theoretical results with measured equivalents, where the feasibility and limitations of such systems have been demonstrated for the first time.

The work of Ms. Sit thus represents an important innovation to the state of the art. I am sure that the OFDM-based radar system will play a substantial role in the future and this work will be in the spotlight and find many applications worldwide. For Ms. Sit, with her creativity and great organizational skills, I wish her success in her scientific and economic endeavors.

Prof. Dr.-Ing. Thomas Zwick
- Institute Director

Forschungsberichte aus dem Institut für Höchstfrequenztechnik und Elektronik (IHE) der Universität Karlsruhe (TH) (ISSN 0942-2935)

Herausgeber: Prof. Dr.-Ing. Dr. h.c. Dr.-Ing. E.h. mult. Werner Wiesbeck

Band 1 Daniel Kähny
Modellierung und meßtechnische Verifikation polarimetrischer, mono- und bistatischer Radarsignaturen und deren Klassifizierung (1992)

Band 2 Eberhardt Heidrich
Theoretische und experimentelle Charakterisierung der polarimetrischen Strahlungs- und Streueigenschaften von Antennen (1992)

Band 3 Thomas Kürner
Charakterisierung digitaler Funksysteme mit einem breitbandigen Wellenausbreitungsmodell (1993)

Band 4 Jürgen Kehrbeck
Mikrowellen-Doppler-Sensor zur Geschwindigkeits- und Wegmessung - System-Modellierung und Verifikation (1993)

Band 5 Christian Bornkessel
Analyse und Optimierung der elektrodynamischen Eigenschaften von EMV-Absorberkammern durch numerische Feldberechnung (1994)

Band 6 Rainer Speck
Hochempfindliche Impedanzmessungen an Supraleiter / Festelektrolyt-Kontakten (1994)

Band 7 Edward Pillai
Derivation of Equivalent Circuits for Multilayer PCB and Chip Package Discontinuities Using Full Wave Models (1995)

Band 8 Dieter J. Cichon
Strahlenoptische Modellierung der Wellenausbreitung in urbanen Mikro- und Pikofunkzellen (1994)

Band 9 Gerd Gottwald
Numerische Analyse konformer Streifenleitungsantennen in mehrlagigen Zylindern mittels der Spektralbereichsmethode (1995)

Band 10 Norbert Geng
Modellierung der Ausbreitung elektromagnetischer Wellen in Funksystemen durch Lösung der parabolischen Approximation der Helmholtz-Gleichung (1996)

Band 11 Torsten C. Becker
Verfahren und Kriterien zur Planung von Gleichwellennetzen für den Digitalen Hörrundfunk DAB (Digital Audio Broadcasting) (1996)

Forschungsberichte aus dem Institut für Höchstfrequenztechnik und Elektronik (IHE) der Universität Karlsruhe (TH) (ISSN 0942-2935)

Forschungsberichte aus dem
Institut für Höchstfrequenztechnik und Elektronik (IHE)
der Universität Karlsruhe (TH) (ISSN 0942-2935)

Band 24 Dirk Didascalou
Ray-Optical Wave Propagation Modelling in Arbitrarily Shaped Tunnels (2000)

Band 25 Hans Rudolf
Increase of Information by Polarimetric Radar Systems (2000)

Band 26 Martin Döttling
Strahlenoptisches Wellenausbreitungsmodell und Systemstudien für den Satellitenmobilfunk (2000)

Band 27 Jens Haala
Analyse von Mikrowellenheizprozessen mittels selbstkonsistenter finiter Integrationsverfahren (2000)

Band 28 Eberhard Gschwendtner
Breitbandige Multifunktionsantennen für den konformen Einbau in Kraftfahrzeuge (2001)

Band 29 Dietmar Löffler
Breitbandige, zylinderkonforme Streifenleitungsantennen für den Einsatz in Kommunikation und Sensorik (2001)

Band 30 Xuemin Huang
Automatic Cell Planning for Mobile Network Design: Optimization Models and Algorithms (2001)

Band 31 Martin Fritzsche
Anwendung von Verfahren der Mustererkennung zur Detektion von Landminen mit Georadaren (2001)

Band 32 Siegfried Ginter
Selbstkonsistente Modellierung der Erhitzung von biologischem Gewebe durch hochintensiven Ultraschall (2002)

Band 33 Young Jin Park
Applications of Photonic Bandgap Structures with Arbitrary Surface Impedance to Luneburg Lenses for Automotive Radar (2002)

Band 34 Alexander Herschlein
Entwicklung numerischer Verfahren zur Feldberechnung konformer Antennen auf Oberflächen höherer Ordnung (2002)

Band 35 Ralph Schertlen
Mikrowellenprozessierung nanotechnologischer Strukturen am Beispiel von Zeolithen (2002)

Forschungsberichte aus dem
Institut für Höchstfrequenztechnik und Elektronik (IHE)
der Universität Karlsruhe (TH) (ISSN 0942-2935)

Band 36 Jürgen von Hagen
Numerical Algorithms for the Solution of Linear Systems of Equations Arising in Computational Electromagnetics (2002)

Band 37 Ying Zhang
Artificial Perfect Magnetic Conductor and its Application to Antennas (2003)

Band 38 Thomas M. Schäfer
Experimentelle und simulative Analyse der Funkwellenausbreitung in Kliniken (2003)

Band 39 Christian Fischer
Multistatisches Radar zur Lokalisierung von Objekten im Boden (2003)

Band 40 Yan C. Venot
Entwicklung und Integration eines Nahbereichsradarsensorsystems bei 76,5 GHz (2004)

Band 41 Christian Waldschmidt
Systemtheoretische und experimentelle Charakterisierung integrierbarer Antennenarrays (2004)

Band 42 Marwan Younis
Digital Beam-Forming for high Resolution Wide Swath Real and Synthetic Aperture Radar (2004)

Band 43 Jürgen Maurer
Strahlenoptisches Kanalmodell für die Fahrzeug-Fahrzeug-Funkkommunikation (2005)

Band 44 Florian Pivit
Multiband-Aperturantennen für Basisstationsanwendungen in rekonfigurierbaren Mobilfunksystemen (2005)

Band 45 Sergey Sevskiy
Multidirektionale logarithmisch-periodische Indoor-Basisstationsantennen (2006)

Band 46 Martin Fritz
Entwurf einer breitbandigen Leistungsendstufe für den Mobilfunk in Low Temperature Cofired Ceramic (2006)

Band 47 Christiane Kuhnert
Systemanalyse von Mehrantennen-Frontends (MIMO) (2006)

Band 48 Marco Liebler
Modellierung der dynamischen Wechselwirkungen von hoch-intensiven Ultraschallfeldern mit Kavitationsblasen (2006)

Forschungsberichte aus dem
Institut für Höchstfrequenztechnik und Elektronik (IHE)
der Universität Karlsruhe (TH) (ISSN 0942-2935)

Band 49 Thomas Dreyer
Systemmodellierung piezoelektrischer Sender zur Erzeugung hochintensiver Ultraschallimpulse für die medizinische Therapie (2006)

Band 50 Stephan Schulteis
Integration von Mehrantennensystemen in kleine mobile Geräte für multimediale Anwendungen (2007)

Band 51 Werner Sörgel
Charakterisierung von Antennen für die Ultra-Wideband-Technik (2007)

Band 52 Reiner Lenz
Hochpräzise, kalibrierte Transponder und Bodenempfänger für satellitengestützte SAR-Missionen (2007)

Band 53 Christoph Schwörer
Monolithisch integrierte HEMT-basierende Frequenzvervielfacher und Mischer oberhalb 100 GHz (2008)

Band 54 Karin Schuler
Intelligente Antennensysteme für Kraftfahrzeug-Nahbereichs-Radar-Sensorik (2007)

Band 55 Christian Römer
Slotted waveguide structures in phased array antennas (2008)

Fortführung als
"Karlsruher Forschungsberichte aus dem Institut für Hochfrequenztechnik und Elektronik" bei KIT Scientific Publishing
(ISSN 1868-4696)

Karlsruher Forschungsberichte aus dem Institut für Hochfrequenztechnik und Elektronik (ISSN 1868-4696)

Herausgeber: Prof. Dr.-Ing. Thomas Zwick

Die Bände sind unter www.ksp.kit.edu als PDF frei verfügbar oder als Druckausgabe bestellbar.

Band 55 Sandra Knörzer
Funkkanalmodellierung für OFDM-Kommunikationssysteme bei Hochgeschwindigkeitszügen (2009)
ISBN 978-3-86644-361-7

Band 56 Thomas Fügen
Richtungsaufgelöste Kanalmodellierung und Systemstudien für Mehrantennensysteme in urbanen Gebieten (2009)
ISBN 978-3-86644-420-1

Band 57 Elena Pancera
Strategies for Time Domain Characterization of UWB Components and Systems (2009)
ISBN 978-3-86644-417-1

Band 58 Jens Timmermann
Systemanalyse und Optimierung der Ultrabreitband-Übertragung (2010)
ISBN 978-3-86644-460-7

Band 59 Juan Pontes
Analysis and Design of Multiple Element Antennas for Urban Communication (2010)
ISBN 978-3-86644-513-0

Band 60 Andreas Lambrecht
True-Time-Delay Beamforming für ultrabreitbandige Systeme hoher Leistung (2010)
ISBN 978-3-86644-522-2

Band 61 Grzegorz Adamiuk
Methoden zur Realisierung von dual-orthogonal, linear polarisierten Antennen für die UWB-Technik (2010)
ISBN 978-3-86644-573-4

Band 62 Jutta Kühn
AlGaN/GaN-HEMT Power Amplifiers with Optimized Power-Added Efficiency for X-Band Applications (2011)
ISBN 978-3-86644-615-1

Karlsruher Forschungsberichte aus dem Institut für Hochfrequenztechnik und Elektronik (ISSN 1868-4696)

Band 63 Małgorzata Janson
Hybride Funkkanalmodellierung für ultrabreitbandige MIMO-Systeme (2011)
ISBN 978-3-86644-639-7

Band 64 Mario Pauli
Dekontaminierung verseuchter Böden durch Mikrowellenheizung (2011)
ISBN 978-3-86644-696-0

Band 65 Thorsten Kayser
Feldtheoretische Modellierung der Materialprozessierung mit Mikrowellen im Durchlaufbetrieb (2011)
ISBN 978-3-86644-719-6

Band 66 Christian Andreas Sturm
Gemeinsame Realisierung von Radar-Sensorik und Funkkommunikation mit OFDM-Signalen (2012)
ISBN 978-3-86644-879-7

Band 67 Huaming Wu
Motion Compensation for Near-Range Synthetic Aperture Radar Applications (2012)
ISBN 978-3-86644-906-0

Band 68 Friederike Brendel
Millimeter-Wave Radio-over-Fiber Links based on Mode-Locked Laser Diodes (2013)
ISBN 978-3-86644-986-2

Band 69 Lars Reichardt
Methodik für den Entwurf von kapazitätsoptimierten Mehrantennensystemen am Fahrzeug (2013)
ISBN 978-3-7315-0047-6

Band 70 Stefan Beer
Methoden und Techniken zur Integration von 122 GHz Antennen in miniaturisierte Radarsensoren (2013)
ISBN 978-3-7315-0051-3

Band 71 Łukasz Zwirełło
Realization Limits of Impulse-Radio UWB Indoor Localization Systems (2013)
ISBN 978-3-7315-0114-5

Band 72 Xuyang Li
Body Matched Antennas for Microwave Medical Applications (2014)
ISBN 978-3-7315-0147-3

Karlsruher Forschungsberichte aus dem Institut für Hochfrequenztechnik und Elektronik (ISSN 1868-4696)

Band 73 Sebastian Diebold
Transistor- und Leitungsmodellierung zum Entwurf von monolithisch integrierten Leistungsverstärkern für den hohen Millimeterwellen-Frequenzbereich (2014)
ISBN 978-3-7315-0161-9

Band 74 Christian Rusch
Integrierte, planare Leckwellenantennen für 3D-Millimeterwellen-Radarsysteme basierend auf dem holografischen Prinzip (2014)
ISBN 978-3-7315-0234-0

Band 75 Marlene Harter
Dreidimensional bildgebendes Radarsystem mit digitaler Strahlformung für industrielle Anwendungen (2014)
ISBN 978-3-7315-0249-4

Band 76 Michael A. Baldauf
Abhängigkeit der Exposition von der Zellgröße beim Mobilfunk unter Gewährleistung der Versorgung (2015)
ISBN 978-3-7315-0308-8

Band 77 Alicja Ossowska
Highly Resolved Synthetic Aperture Radar with Beam Steering (2015)
ISBN 978-3-7315-0315-6

Band 78 Małgorzata Dominika Brzeska
RF Modelling and Characterization of Tyre Pressure Sensors and Vehicle Access Systems (2015)
ISBN 978-3-7315-0348-4

Band 79 Ulrich Lewark
Aktive Frequenzvervielfacher zur Signalerzeugung im Millimeter- und Submillimeterwellen Frequenzbereich (2015)
ISBN 978-3-7315-0354-5

Band 80 Kai-Philipp Walter Pahl
Distributed Transformers for Broadband Monolithic Millimeter-Wave Integrated Power Amplifiers (2015)
ISBN 978-3-7315-0409-2

Band 81 Serdal Ayhan
Hochgenaue radarbasierte Abstandsmessung mit geführter Wellenausbreitung (2016)
ISBN 978-3-7315-0433-7

Karlsruher Forschungsberichte aus dem
Institut für Hochfrequenztechnik und Elektronik
(ISSN 1868-4696)

Band 82 Yoke Leen Sit
MIMO OFDM Radar-Communication System with Mutual Interference Cancellation (2017)
ISBN 978-3-7315-0599-0

MIMO OFDM Radar-Communication System with Mutual Interference Cancellation

Zur Erlangung des akademischen Grades eines

DOKTOR-INGENIEURS

von der Fakultät für
Elektrotechnik und Informationstechnik
des Karlsruher Instituts für Technologie (KIT)

genehmigte

DISSERTATION

von

M. Sc. Yoke Leen Sit
geb. in Penang, Malaysia

Tag der mündlichen Prüfung:	27.07.2016
Hauptreferent:	Prof. Dr.-Ing. Thomas Zwick
Korreferent:	Prof. Dr.-Ing. Christian Waldschmidt

Acknowledgement

Two roads diverged in a wood, and I -
I took the one less traveled by,
And that has made all the difference.

- Robert Frost, *The Road Not Taken*

My heartfelt gratitude goes to the many people who have helped me in the pursuit of a PhD. I have learnt much from others, helped raised a new generation of curious scientist-engineers, and am humbled by them all. With this, I close another significant chapter of my life. A new adventure awaits.

Prime motivator:	Prof.Dr.-Ing. Thomas Zwick
OFDM Radar team:	Dr.-Ing. Christian Sturm, Dr.-Ing. Martin Braun, Benjamin Nuß
Antenna & USRP team:	Jerzy Kowalewski, Tobias Mahler
Students:	Johannes Baier, Thanh Thuy Nguyen, Hao Liu, Sanjoy Basak, Mateusz Orzol, Dorota Truchanowicz
Technicians:	Michael Harrer, Andreas Lipp, Mirko Nonnenmacher, Andreas Gallego, Ronald Vester
Proofreaders:	Dr.-Ing. Christian Sturm, Benjamin Nuss, Dr.-Ing. Lukasz Zwirello, Dr.-Ing. Xuyang Li
Collaborators:	Dr. Bita Sobhani, Dr. -Ing. Lars Reichardt, Dr.-Ing. Tom Schipper, Dr.-Ing. Marlene Harter, Christoph Heine, Benjamin Göttel, Prof. Dr.-Ing. Dr.h.c. Dr.-Ing. E.h. mult. Werner Wiesbeck
Secretaries:	Simone Gorré, Angela Ziemba
Supporters:	IHE colleagues, family and friends

Yoke Leen Sit
Autumn 2016

Abstract

While the radar is not a new concept, the evolution of radars is very slow compared to communication technologies. Ever since the rapid development of consumer electronics, especially the introduction of radars in the automotive sector, the radar revolution began.

The trend for the automotive sector is currently driven by new requirements defined for future vehicles with safer driving and convenience as a major aspect. As part of the sustainable transportation plan for the future, the European Union has been very active in pursuing vehicle-to-vehicle (V2V) communication systems within the plans for intelligent transportation systems (ITS). With ever increasing traffic densities, connectivity between vehicles and even infrastructures (V2I) can make navigation a collaborative effort rather than an individual one. The concept of collaborative driving ultimately boils down to the interaction between the multiple sensors and the communication system that will determine the decisions made by the learning algorithms, which in turn translate it into a reaction. However, as long as the intention of other drivers are not known or communicated an accident prevention maneuver cannot be premeditated.

Herein lies the proposal of this work - a multiple-input multiple-output (MIMO) OFDM-based joint radar-communication (RadCom) system, where the MIMO radar feature does not require a separate set of hardware. This system would hence be highly efficient in terms of spectrum, hardware, size, and will eventually also be cost effective. The signal model uses the OFDM signal, which is detailed in this thesis, from the generation of the signal to host multiple antennas as well as users to the radar signal processing for the range, azimuth, elevation and Doppler. Since the antenna arrays and signal model also influences the signal processing and accuracy of the radar estimation, research has also been dedicated to these two topics.

For the operation in a network, it is inevitable that multiple RadCom nodes will attempt to perform both communication and radar imaging at the same time. Although this is somewhat mitigated by allowing each node to transmit only on selected non-overlapping channels, OFDM, being notoriously susceptible to subcarrier misalignment will without fail experience mutual interference in a real and non-ideal scenario. This work hence analyzes the mutual interference tolerance of a MIMO OFDM RadCom showing that in a regular automotive scenario, the hardware imperfection and movement of the radar, interferer or targets will definitely lead to interference to the radar.

As such an interference cancellation algorithm, especially one that is simple in implementation yet effective will be necessary. Interference cancellation for OFDM-based radar is unlike the one for OFDM communication systems. The idea of interference cancellation in communication systems is to remove enough interference to recover the payload data. For the radar however, its goal is to separate the entire received interferer signal from the desired useful signal that is reflected from the targets. The better the separation and subsequent cancellation of the interfering signal, the higher the signal-to-noise and interference ratio (SNIR) of the radar measurement. That being said, the process of separating the desired and undesired signals at the radar's receiver is not straightforward. While regular OFDM demodulation techniques used in communication systems can be applied to the radar as well, these techniques have insufficient accuracy to correctly reproduce the interfering signal to be subtracted from the overall received signal. Hence additional steps must be taken to narrow the parameter deviation and these steps have been presented in detail in the thesis along with three case studies that explain and verify the interference cancellation algorithm.

To verify the system concept of the MIMO RadCom as well as the effectiveness of the interference cancellation algorithm, the system is implemented on universal software radio peripherals (USRPs) with a scaled-down operating frequency of $f_c = 4.05\,\text{GHz}$ to take hardware imperfections and propagation losses into account. The USRPs were configured and calibrated to host four synchronized transmitters and receivers to operate in real time with an instantaneous bandwidth of 100 MHz. The system has been successfully tested and verified in indoor and outdoor scenarios with very promising results.

Zusammenfassung

Obwohl das Konzept hinter Radar nicht neu, schreitet die Entwicklung der Radartechnologie sehr langsam voran im Vergleich zur Kommunikationstechnik. Seitdem jedoch immer mehr Radare im Automobilbereich Anwendung finden, nimmt die Revolution auf dem Radarmarkt langsam Fahrt auf.

Der gegenwärtige Trend für den Automobilbereich ist getrieben durch neue Anforderungen, die sich durch zukünftige Fahrzeuge ergeben, wie z.B. neue Sicherheitsaspekte und steigender Komfort. Als Teil des zukünftigen Mobilitätskonzepts hat die Europäische Union unter anderem Vehicle-to-Vehicle (V2V) Kommunikationssysteme ausgemacht, die wiederum Teil eines Intelligenten Transportsystems (ITS) sein sollen. Mit steigenden Verkehrszahlen stellt V2V in Verbindung mit einer Kommunikation zwischen Fahrzeug und Infrastruktur (V2I) eine Möglichkeit dar, Navigation in Zukunft im Verbund zu lösen und nicht mehr individuell in jedem einzelnen Fahrzeug. Das Konzept des vernetzten Fahrens setzt hierbei bereits bei der Interaktion zwischen den Sensoren und den Kommunikationssystemen an, die die Entscheidungen von selbst lernenden Algorithmen beeinflussen, die wiederum schließlich eine Handlungsentscheidung treffen. Solange jedoch die Absichten der einzelnen Fahrzeuge nicht bekannt sind oder übermittelt werden, kann ein Maneuver zum Verhindern von Unfällen nicht zuverlässig arbeiten.

Hierin liegt einer der Kernpunkte der vorliegenden Arbeit: ein auf Orthogonal Frequency Division Multiplexing (OFDM) basierendes Multiple Input Multiple Output (MIMO) System, das sowohl Radar als auch Kommunikation vereint. Einer der Vorteile liegt dabei darin, dass für die Kommunikation keine separate Hardware benötigt wird, wodurch das Gesamtsystem sehr effizient im Bezug auf Spektrum, Hardwareeinsatz, Größe und eventuell sogar im Hinblick auf Materialkosten ist. Das Signalmodell basiert auf OFDM und wird in dieser Arbeit detailliert beschrieben - angefangen von der Erzeugung des Sendesignals, der Zuteilung der Signale zu den Antennen sowie der verschiedenen Nutzer auf unterschiedliche Subkanäle bis hin zur Schätzung von Entfernung, Winkel und Geschwindigkeit der Ziele. Da die Antennenarrays sowie das Signalmodell einen großen Einfluss auf die Signalverarbeitung und die Schätzgenauigkeit haben, wurden auch diese beiden Themen gründlich untersucht. In einem Netzwerk werden zwangsläufig mehrere RadCom-Anwender gleichzeitig versuchen, Radar als auch Kommunikation einzusetzen. Obwohl Störung -en durch den gleichzeitigen Betrieb teilweise dadurch vermieden werden, dass jeder beteiligte RadCom-Knoten nur ein nicht überlappender Teil der Unterträger zugewiesen bekommt, reagiert OFDM sehr empfindlich auf eine Verschiebung dieser Unterträger. Im schlimmsten Fall kann diese Art der Interferenz zu einem Totalausfall des Radarsystems führen. Aus diesem Grund wird in dieser Arbeit die Toleranz des vorgeschlagenen MIMO OFDM RadCom Systems hinsichtlich dieser gegenseitigen Störungen untersucht. Hierbei wird gezeigt, dass in einem alltäglichen automobilen Szenario die nicht perfekte Hardware, die Bewegung des Radars, der Störer sowie der Ziele definitiv zu einer Beeinträchtigung des Radars führen.

Aus diesem Grund wird ein Algorithmus zur Interferenzunterdrückung benötigt, der sowohl einfach in der Implementierung als auch sparsam im Rechenaufwand ist. Unglücklicherweise lässt sich die Interferenzunterdrückung nicht so einfach bewerkstelligen wie bei OFDM basierten Kommunikationssystemen. Bei diesen genügt es, einen ausreichend hohen Anteil der Interferenz zu entfernen, um gerade so die Nutzdaten wiederherzustellen. Beim Radar hingegen ist die Zielsetzung, das gewünschte, vom Ziel reflektierte Signal vollständig aus dem Empfangssignal zu extrahieren. Je besser die Unterdrückung der Interferenz gelingt, desto höher ist das Signal-zu-Rausch-und-Interfernz-Verhält -nis (SNIR) der Radarmessung. Die hierfür notwendigen Arbeitsschritte sind jedoch nicht direkt ersichtlich. Während die gewöhnlichen Demodulationstechniken für OFDM, die in Kommunikationssystemen angewendet werden, zwar auch beim Radar angewendet werden können, sind deren Ergebnisse bei genauerer Betrachtung jedoch zu ungenau, um das Interferenzsignal zu rekonstruieren und letztlich vom Empfangssignal zu subtrahieren. Aus diesem Grund sind zusätzliche Schritte notwendig, um die verbleibenden Parameterabweichungen zu korrigieren. Die notwendigen Schritte sind in dieser Arbeit im Detail erläutert zusammen mit drei Fallbeispielen, die die Wirksamkeit des entwickelten Algorithmus aufzeigen. Um das Konzept des MIMO RadCom Systems sowie die Effektivität des Algorithmus zur Interferenzunterdrückung zu verifizieren, wurde das System auf Universal Software Radio Peripherals (USRPs) implementiert, wodurch alle Einflüsse der Hardware sowie Verluste ebenfalls sichtbar wurden. Bedingt durch die Hardware war es dabei nötigt die Trägerfrequenz auf $f_c = 4.05\,\mathrm{GHz}$ herunterzuskalieren. Die USRPs wurden so konfiguriert und kalibriert, dass vier Sender und Empfänger gleichzeitig in Echtzeit mit einer instantanen Bandbreite von 100 MHz betrieben werden konnten. Das System wurde erfolgreich getestet und verifiziert und Messungen unter realen Bedingungen im Freien haben vielversprechende Ergebnisse geliefert.

Table of contents

Acronyms and symbols

Lower case letters

c_0	speed of light in vacuum ($\approx 2.997925 \cdot 10^8$ m/s)
$\vec{d}_{\mathrm{R}_q}$	vector of the q-th receive antenna element to the origin
d_T	transmit antenna array element spacing
$\vec{d}_{\mathrm{T}_p}$	vector of the p-th transmit antenna element to the origin
d_R	receive antenna array element spacing
f_c	center frequency or frequency of operation
$f_{\mathrm{D},h}$	Doppler frequency of target h
$f_{\mathrm{D,max}}$	maximum Doppler frequency
h	target index
i	counter for subcarrier indices associated with u
m	OFDM symbol index
n	OFDM subcarrier index
n_u	OFDM subcarrier indices allocated to the user index u
p	transmit antenna index
q	receive antenna index
r_h	distance of target h from the radar
$r_{p,h}$	distance of target h from the p-th transmit antenna
$r_{q,h}$	distance of target h from the q-th receive antenna
u	user index
v_rel	relative velocity
$\mathbf{v}_\mathrm{T}$	transmit beamsteering vector
$\mathbf{v}_\mathrm{R}$	receive beamsteering vector

(x,y,z)	Cartesian coordinates
x_{T_p}	x Cartesian coordinate of the p-th transmitter
x_{R_q}	x Cartesian coordinate of the q-th receiver
$\mathbf{x}_{\mathrm{T}_p}$	(x,y,z) Cartesian coordinate of the p-th transmitter
$\mathbf{x}_{\mathrm{R}_q}$	(x,y,z) Cartesian coordinate of the q-th receiver

Capital letters

$\mathcal{A}$	modulation alphabet
BW	signal bandwidth
DFT	discrete Fourier transform
D	radar quotient element
$\mathbf{D}$	radar quotient matrix
$\mathbf{FB}$	Fourier Beamforming matrix
$\mathcal{F}$	Fourier transformation
$\mathcal{F}^{-1}$	inverse Fourier transformation
G	gain or loss factor
G_{per}	processing gain from the periodogram method
$G_{\mathrm{T}}, G_{\mathrm{R}}$	transmit-, receive antenna gain
H	number of targets
$\mathbf{H}$	channel transfer function
$\mathbf{I}$	processed radar matrix
IDFT	inverse discrete Fourier transform
L	path loss
N	number of subcarriers
N_{ch}	number of user channels
N_{CP}	number of cyclic prefix symbols
N_{f}	pilot tone spacing in the frequency-subcarrier axis
N_{t}	pilot tone spacing in the time-symbol axis
N_{train}	number of training symbols
M	number of contiguous OFDM symbols without training symbols
$\mathcal{P}$	pilot tone matrix
P	number of transmit antennas
$P_{\mathrm{T}}, P_{\mathrm{R}}$	transmit-, receive power
Q	number of receive antennas

$\mathbb{R}$	real number
$\mathbf{S}$	Comm transmit matrix
T_0	OFDM symbol duration
T_{CP}	duration of guard interval or cyclic prefix
T_{S}	sampling time
T	total duration of one OFDM symbol with cyclic prefix
$\mathbf{W}$	Comm receive matrix
$\mathbf{X}_p$	radar transmit matrix at the p-th transmitter
$\mathbf{Y}_{qp}$	radar receive matrix at the q-th receiver from the p-th transmitter
Z	(Gaussian white) noise term
$\mathbf{Z}$	(Gaussian white) noise matrix
$\mathbb{Z}$	integer number

Mathematical operators and indexes

$\square^{\mathrm{T}}$	transpose matrix
$\square^{*}$	conjugate matrix
$\|\cdot\|$	Euclidean norm of vector
$*$	convolution
$\in$	subset value or element of
$\circ$	Hadamard product
$\otimes$	Kronecker product
$\vec{\square}$	vector
$\square_w$	variable associated with Comm (mutual interferer)
$\hat{\square}$ or $\tilde{\square}$	estimated value
$\{\cdots\}$	range or sequence of values
$[a,b]$	values confined within the range and inclusive of a and b
$\lfloor a \rfloor = b$	floor function where $\lfloor a \rfloor = \max\{b \in \mathbb{Z} \mid b < a\}$
$\therefore$	therefore

Greek symbols

α_h	attenuation due to the path of target h
Δf	subcarrier spacing
Δf_{D}	Doppler resolution
Δr	radar range resolution
δ_w	total frequency offset of Comm
σ_h	radar cross section of target h
τ_h	time delay of path due to target h
θ_h	elevation angle of target h
$\vartheta_{\mathrm{HPBW}}$	half power beam width
κ_{r_h}	range phase rotation term of target h
κ_{D_h}	Doppler phase rotation term of target h
λ_{c}	operating wavelength
Ψ_h	phase offset due to two times the Euclidean distance from target h to radar
ϕ_h	azimuth angle of target h

Acronyms

1D, 2D, 3D, 4D	One-, Two-, Three-, Four-dimensional
ADC	Analog to Digital Converter
AF	Array Factor
AWGN	Additive White Gaussian Noise
BER	Bit Error Ratio
BPF	Bandpass Filter
CFAR	Constant False Alarm Rate
CFO	Carrier Frequency Offset
Comm	Communication partner and potential mutual interferer of the same RadCom system
CP	Cyclic Prefix
DAC	Digital to Analog Converter
DC	Direct Current
DOA	Direction of Arrival

DR	Dynamic Range
Eq.	Equation
EU	European Union
Fig.	Figure
FMCW	Frequency Modulated Continuous Wave
FPGA	Field Programmable Gate Array
GI	Guard Interval
GPS	Global Positioning System
HPBW	Half-Power Beamwidth
ICI	Inter Carrier Interference
IF	Intermediate Frequency
I&Q	In-Phase and Quadrature
ISI	Inter Symbol Interference
LIDAR	Light Detection and Ranging
LNA	Low Noise Amplifier
LO	Local Oscillator
LOS	Line of Sight
LPF	Lowpass Filter
MIMO	Multiple Input Multiple Output
MUSIC	MUltiple SIgnal Classification Algorithm
NLOS	Non Line of Sight
NWA	Network Analyzer
OFDM	Orthogonal Frequency Division Multiplexing
PA	Power Amplifier
PAPR	Peak-to-Average Power Ratio
PLL	Phase Locked Loop
PN	Pseudo Noise
PSK	Phase Shift Keying
QAM	Quadrature Amplitude Modulation
Radar	Ego RadCom node under analysis
RadCom	Radar-Communication System
RCS	Radar Cross Section
RF	Radio Frequency
Rx	Receiver
SCA	Schmidl and Cox Algorithm
SDR	Software Defined Radio

SDRS	Software Defined Radio System
SIR	Signal to Interference Ratio
SISO	Single Input Single Output
SNIR	Signal to Noise and Interference Ratio
SNR	Signal to Noise Ratio
TDM	Time Division Multiplexing
Tx	Transmitter
UAV	Unmanned Aerial Vehicle
ULA	Uniform Linear Array
USRP	Universal Software Radio Peripheral
V2V	Vehicle-to-Vehicle
V2V	Vehicle-to-Infrastructure
VCO	Voltage Controlled Oscillator
WLAN	Wireless Local Area Network
XCORR	Cross-correlation

All other specific abbreviations will be introduced in the text.

1 Introduction

There will always come a time when even the best and most conventional of things is reinvented. It is somewhat like trying to reinvent the wheel except that wings were instead discovered. These new reinventions, coexisting along with the conventional, will offer more options and variety to push forward the current age of creative technical boom. The radar is currently one such technology for reinvention.

Radars were first developed for warfare to detect aircraft and ships. Their evolution has advanced rapidly after World War II and they are now used in many civilian or commercial applications from navigation, precision measurement, to area surveillance. As a continuum to traditional radars, came 'networked' and'multiple input multiple output' (MIMO) radars, both of which are garnering a great deal of interest.

The concept of networked radars is not new. Any radar that is not part of the same hardware platform and are located separately yet must collaborate (whether cooperatively or non-cooperatively) to yield some form of estimation is in essence, a form of networking. One of the first such operational radars was the passive bistatic system developed during World War II called the *Klein Heidelberg* that used British Chain Home signals as illuminators [GW10]. It was an ingenious piece of technology that was decades ahead of its time[1] that evaded the scrutiny of the British but was deployed too late to have much effect on the outcome of the war. Since then many passive bistatic and multistatic have been developed and refined for the same purpose of surveillance. One of most recent development was to use WiFi signals as illuminators of opportunity for indoor and outdoor localization as proposed by Falcone et al. in [FCL12].

[1] covered rather extensively by Bauer in [khr10]

MIMO radars on the other hand, are relatively new with origins in the communication technology. While communication technologies are concerned with getting information across the link, the radar's primary interest is to gauge the channel. The concept of MIMO used on radars was hailed as a game changer by [FHB+04, LS08]. At the same time it is also met with much skepticism especially by industry veterans such as Brookner from Raytheon in [Bro15, Bro13] and Daum in [DH09] who maintained that the conventional phased array radars still outperform most of the claims of advantages of the MIMO radar due to unfair comparisons made. These claims touch mostly on the size of the main lobe generated by the full or sparse antenna array, which in turn directly influences parameter identifiability of targets; as well as the complexity and processing power required by MIMO radars, all of which are prominent factors required of military or defense radars.

It must be stressed here that conventional phased array radars have been around much longer than its MIMO radar counterpart and hence its technology is also much more mature. MIMO radars are only starting to find its footing with almost all researches to date concentrating only theoretical developments or simulations. What can be conceded however, is that MIMO radars and phased array radars will most likely be employed for different applications, with the phased array in the mid- to long-range surveillance and target tracking in aerospace and defense, while MIMO radar will find its niche in consumer and commercial electronics.

With the advent of modularized electronic units, advances in computational processing power, ideas that once confined the MIMO radar concepts to mere theories can now be brought forth and realized. These modularized hardware consisting of ever compacter system-on-chips (SOC) enables cheaper manufacturing costs while cramming more features within - the current primary factor that determines the success of a commercial electronic product. The phased array radar with all its superior tracking and surveillance abilities is usually massive in size and cannot be miniaturized without the severe repercussions of decreased radar capabilities.

To see the comparison, consider here two uniform linear array antennas both in the azimuth and elevation planes with P and Q being the number of transmit and receive antennas respectively. If a phased array radar were to be compared to the MIMO radar to result in the same angular resolution, the phased array radar will require $P \cdot Q$ number of antennas as compared to $P + Q$ antennas for the MIMO radar case. Even if one could argue that the signal processing hardware and effort is less for the case of the phased array, this advantage is not immediately apparent when the number of antennas used is small. Considering this simple argument the phased array radar is foreseen to remain within its current market and applications.

The major leaders of radar development are in the automotive and security industries and this in turn affects the direction of radar concept developments. The current trend in the automotive industry is autonomous driving assistance for better road safety. This is evident in the ambitious development of self-driving cars by Tesla Motors and Google as well as up and coming cooperative projects by various universities and industries in the European Union (EU). Meanwhile with the sudden rise of the use of unmanned aerial vehicles (UAV) especially drones, the development of active area surveillance radars is also on the cards. The aforementioned trends will be explored in the following sections.

1.1 Radar networks

1.1.1 Vehicular radar networks

The trend for the automotive sector is currently driven by new requirements defined for future vehicles with safer driving and convenience as a major aspect. The statistics from the World Health Organization (WHO) show that road injury is the leading cause of death after major diseases [who20b], with a significantly higher death toll in low to middle income countries, while higher income countries have seen a decrease in the number of death. This comes despite the low to middle income countries having just 54 % of the world's vehicles yet see 90 % of road fatalities [who20a]. It can be surmised that road safety laws and vehicle standard regulations mandated by high income countries play a crucial role in this matter.

As part of the sustainable transportation plan for the future, the EU has been very active in pursuing vehicle-to-vehicle (V2V) communication systems within the plans for intelligent transportation systems (ITS) under the DRIVEC2X project in [dri10] and MOSARIM project in [mos20]. With ever increasing traffic densities, connectivity between vehicles and even infrastructures (V2I) can make navigation a collaborative effort rather than an individual one. To observe how this collaborative effort might provide better driving assistance and overall road safety, the development of autonomous cars can be used as example case studies.

One of the most popular semi-autonomous or autopilot cars in the current market is Tesla Motor's Model S. The car is equipped with "a forward radar, a forward-looking camera, 12 long-range ultrasonic sensors positioned to sense 5 m around the car in every direction at all speeds, and a high-precision digitally-controlled electric assist braking syste" as mentioned in [tes20b]. This autopilot feature relies heavily on all its sensors and learning algorithms allowing it to steer safely within a straight road, initiate lane change, manage speed and even park the car. Nevertheless it does not enable a true autonomous mode, whereby the driver exert minimal to no control over the navigation mainly because the system is not yet intelligent enough to devise a safe reaction toward more complicated traffic scenarios. This situation is also true for all the other semi-autonomous cars such as the BMW 750i, Infinity Q50S and Mercedes Benz S65 AMG as tested in [sem20].

On the other hand, there is Google's highly ambitious self-driving car, with the interior designed solely for riding, not driving. Such an attempt would of course require even more sophisticated sensors, namely the Velodyne LIDAR (initially costing USD 75,000), other 360° radars, cameras and a computer with learning algorithm and pre-loaded knowledge of the area such as speed limits. While Google claimed to have successfully tested the car in many rigorously complicated driving scenarios, a very recent event on the 14th of February 2016 saw the Google car in Mountain View California attempting a maneuver to avoid some sandbags while making a right turn but struck an oncoming bus instead. The fault lies in the decision making algorithm, or the lack of it, for assuming that the oncoming bus would slow down when in fact larger vehicles are less likely to yield. This points to the crucial lack of human judgment that even if can be programmed into the decision making algorithm (requiring a large database and processing power), might not be applicable universally.

In it precisely the case from the two autonomous car examples given that a vehicular network would be beneficial. At the end of the day, the aim of these autonomous car makers is to put as many autonomous cars as possible on the road. Instead of having all vehicles navigate independently, having them communicate and collaborate their navigation and sensor data (through V2V and V2I) could potentially ease the large amount of data to be stored and processed by any one autonomous car and at the same time make collision avoidance and traffic flow a multi-vehicle collaborative effort. Tesla Motors already has plans to place charging stations across the USA hence building networking infrastructure to aid navigation along roads is not too far fetched an idea either [tes20a].

1.1.2 Networked radars for area surveillance

In the same way for area surveillance radars, a network consisting of multiple cooperative radar nodes will be able to exchange information among themselves to reduce blind spots and perform data fusion to obtain a higher quality radar estimates. The major difference of this application compared to the vehicular network is that the radar nodes tend to be static. The advantage of using a radar network instead of a solitary radar is manifold. toward the detection of targets with very small radar cross section (RCS), or stealth targets that are shaped to scatter energy away from the direction of the incident wave, as well as targets with erratic paths, i.e. UAV or drones, a multistatic structure can better capture the scattered energy from the target since it is more unlikely to be obsured over all different transmit and receive paths. It has also been shown in [Sie58] that the forward scatter of a small RCS near to the bistatic baseline can be substantially higher than the monostatic RCS though it comes with a penalty of poor range and Doppler resolution.

Most of such multistatic radars are passive in nature and are used to observe weather patterns such as in [GG03, WRF+94] or developed for defense systems (available from Cassidian, Thales and Lockheed Martin) to exploit the already available transmitters of opportunity and the relatively low cost of the receivers. For a commercial application, an active networked radar would be more rational for use. Take for example the increasing need to survey public areas for threats, such as a drone (which might potentially be used to transport weapons or cameras). Several portable and innocuous-looking radar nodes can be placed to surround the area of observation and networked to raise the alarm when a possible threat is detected by any one node or through the decision made by observing the target(s)' behavior.

1.1.3 Where does MIMO radar fit in?

Thus far, the overview only touched on radar networks but not on the radar node itself. This is where the MIMO radar configuration comes in. Even though

'MIMO radar' is the current buzzword, there is no clear definition yet but it is widely understood to be any radar that performs parallel transmission and reception of independent signals. While there are collocated and widely distributed MIMO radars mentioned in literature, only the former will be considered here for simplicity but the concept can nevertheless be extended.

The main purpose of using a closely-spaced antenna array with the transmitter and receiver ideally on the same platform is to add extra dimensions to the radar's perception. A conventional single transmit-receive pair (also called single input single output, SISO) radar configuration is only able to estimate range and Doppler[2]. By adding an antenna array in the azimuth (horizontal) and elevation (vertical) planes, a three-dimensional (3D) estimation in space can be performed. The azimuth and elevation are estimated in terms of direction of arrival (DOA). The velocity[3] is not a spatial dimension, hence if the radar node can also perform velocity estimation, it is said to be able to estimate the 3D+velocity parameters.

The key criterion of a MIMO radar is the use of independent (or orthogonal) transmit signals, analogous to its MIMO communication counterpart. Hence many code-division multiple access (CDMA) based waveforms became the potential candidates. They ideally should possess a high autocorrelation feature but low cross-correlation with other signals, but such perfection in practice is difficult to achieve. As such, many early MIMO radar pioneers concerned their research with analyzing potential waveforms such as Bliss et al. in [BFD$^+$09] and Li and Stoica in [LSX06, HSL09]. Processing these signals would require as many matched filter banks as the number of independent transmit signals, making real-time implementation impractical.

In the same way as with CDMA signals, the orthogonal frequency division multiplexing (OFDM) modulation, typically used for communication is also considered. The use of OFDM for radar was first suggested by Levanon in [Lev00], then further researched in Franken et al. in [FNvG06]. Garmatyuk et al. then described its usage for both radar and communication in [GSKS09] but failed to describe the signal processing thus giving little clue to how the radar estimation is performed.

2 range and Doppler detection capabilities are dependent on the waveform used

3 radial velocity component of the relative movement that can be measured from the Doppler shift

Sturm and Wiesbeck in [SPZW09, SW11] were the first to suggest a simple periodogram method of signal processing implementable only with Fast Fourier Transforms for the OFDM radar where any arbitrary data (or payload data) can be used as the radar signal. The radar estimation is therefore independent of the data and in that way, radar sensing and communication can be performed in parallel. Together with Braun in [SZWB10, SBZW10, BSNJ09], experimental proofs of the OFDM signal viability for radar and communication (RadCom) have been presented. From henceforth the concept of a dual-use radar and communication system became one step closer to reality.

A MIMO RadCom is envisioned to have the capability to perform the aforementioned 3D+velocity parameter estimation with the payload data piggybacked on the radar signals. Such a system would be highly efficient in terms of spectrum, hardware, size, and eventually also be cost effective. As have been discussed in the vehicular radar network section, the current so-called autonomous car features multiple sensors. With a normal car already embedded with about 50 antennas for various services[4] with antennas that must be placed accordingly to their functions for optimal performance [RFZ09], the available space within the housing of a car to cram even more services (without compromising the car's aesthetics) is currently very limited. All things considered, the MIMO RadCom is one potential solution that can see the reduction of several different automotive radars to one system. The communication feature then adds to the V2V and V2I concept, further reducing the need for a separate system, a point that has been corroborated by Reichardt et al. using the system presented by Sturm in [RSGZ12].

1.2 Motivation and goals

With OFDM gaining prominence within the radar research circle and the advantages of a RadCom outlined in the previous section, this thesis seeks to design a MIMO RadCom system concept by expanding on the previous groundwork of SISO OFDM RadCom done by Sturm [Stu12] and Braun [Bra14]. Both Sturm and Braun had envisioned a networked-radar application scenario, whereby the signal model must cater to multiple users, since the real capabilities of OFDM-based radar systems only become clear when operated within a network to provide services as stipulated by the ITS concept. This is because

4 AM/FM/DAB radio, GPS, Bluetooth, collision avoidance, etc. [car20].

in terms of range and Doppler detection only, the performance of SISO OFDM radar is equal to that of conventional radars such as the frequency modulated continuous wave (FMCW) radar. Hence a single OFDM radar cannot offer much advantages (in terms of radar sensing only) beyond what conventional radars can already do. Nevertheless, the capability of OFDM radars for dual sensing and communication remains its biggest potential.

This work done in this thesis is divided into two major parts. The first part deals with the individual MIMO RadCom system comprising the signal model, antenna array and radar processing at the receiver as depicted in Fig. 1.1. The second part then explores the interaction between two MIMO RadCom nodes - the ego node 'Radar' and the interferer node 'Comm'. Taking into account the environment with moving targets and nodes as well as hardware imperfection of the RadCom system, an efficient mutual interference cancellation algorithm is proposed so that the MIMO RadCom system will be able to function in a real scenario. The major works done in this thesis are grouped and highlighted together with their motivation and goals as follows.

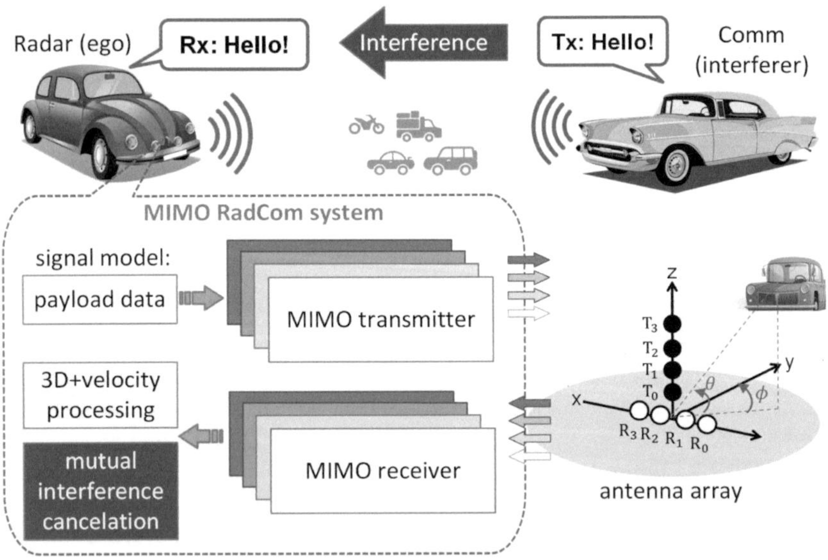

Fig. 1.1: Visualization of the MIMO RadCom system and interaction between RadCom nodes in a typical automotive scenario.

I. MIMO RadCom concept - signal model and virtual antennas

Sturm did a preliminary investigation on the multiple user access scheme, which he called the 'spectrally interleaved' OFDM signal model in [SSBZ13, ZS13]. In this signal scheme, each user will be allocated non-overlapping sets of subcarriers spanning the whole signal bandwidth to use. In this way, the range resolution will not be compromised. The investigation of the channel isolation showed that it can be as high as 70 dB even with some frequency offset. What has not yet been proven however is the applicability of this spectrally-interleaved scheme for DOA estimation.

It has been pointed out by Sen and Nehorai in [SN09] that due to its different frequencies, OFDM signals can resonate better with the frequency-dependent scattering centers of a target to provide richer information [Wei10, Jun04]. In the same way as Wu et al. in [WKG10], the authors only took advantage of the orthogonal nature of the OFDM waveform. Sen and Nehorai used a mutual information-based criterion while Wu et al. used random permutated symbols that would result in a low peak-to-average ratio (PAPR). Both of these techniques were done with hardware limitations in mind, especially the limited bandwidth of the analog/digital converters, which also means that these signals carry no payload data.

As such, the first goal of this work is a thorough analysis of the spectrally interleaved OFDM signal model and its effects on the radar estimation accuracy, as will be presented in Chapter 2. It was found that the user index, which is used as a means to allocate the non-overlapping subcarriers to the multiple-antennas or multiple-users does affect the DOA estimation accuracy.

In order to set up the DOA estimation system, it is also necessary to take the antenna arrays into account. Chapter 3 aims to provide an insight to the concept of 'virtual array', a technique that is frequently used in MIMO radars to maximize its antenna baseline with minimal physical antenna elements. This technique is not applicable to conventional phased array radars, hence the envisioned smaller size of MIMO radars.

II. Mutual interference mitigation and interference cancellation algorithm

For the operation in a network, it is inevitable that multiple RadCom nodes will attempt to perform both communication and radar imaging at the same time. Although this is somewhat mitigated by allowing each node to transmit only on selected non-overlapping channels, OFDM, being notoriously susceptible to subcarrier misalignment will without fail experience mutual (or inter-system) interference in a real and non-ideal scenario.

Before 2013, there was still no metric applicable to gauge OFDM radar networks. Braun investigated the scenario of a networked radar and came up with a new performance metric called the *Radar Network Outage* in [BTJ13]. This outage is defined as the case when the reflected power at the radar's receiver drops below a certain threshold due to the interference created by randomly located nodes (which use either the same or different systems but at the same frequency band). With a radar as proposed by Sturm in [SPZW09], the probability of an outage is calculated to be less than 1% for traffic densities below one node per 158.5 m^2.

In a regular city or highway traffic scenario, the quoted traffic density of one node per 158.5 m^2 is highly underestimated. This leads to the notion that besides interference mitigation, some form of interference cancellation will definitely be required. The goal of Chapter 4 is therefore to investigate the nature of a communication (Comm) partner RadCom node that can potentially turn into an interferer, with the main cause of subcarrier misalignment being the Doppler shift and the carrier frequency offset (CFO) due to imperfect hardware. With all the aforementioned influences taken into consideration, the tolerance of the MIMO RadCom toward such interference is charted and analyzed. Chapter 5 then proposes a flexible and low-complexity interference cancellation algorithm along with several case studies featuring MIMO node interferers. The flexibility and simplicity are the keys for its implementation in real-time operations and scenarios.

III. Subcarrier allocation and demonstration of tracking capabilities

Chapter 6 serves to present the MIMO RadCom system as a whole by taking into considerations the signal model presented in Chapter 2 and the influence of the antennas in Chapter 3. A subcarrier allocation strategy along with the correct choice of antenna elements positions is presented along with their justifications. It has been seen that a less than optimal allocation strategy can affect the target estimation rendering it ambiguous, hence the importance.

Since the motivation of this thesis also alluded to the MIMO RadCom's application for area surveillance applications, a simple toy radar using parameters suitable for implementation using commercially available Software Defined Radios is presented. This toy radar simulation encompass a simple tracking algorithm at its post-processing to simultaneously detect and distinguish the multiple targets by tracking their trajectories. When several of these RadCom nodes are positioned to surround an area of observation, new data fusion techniques as recently proposed by Nuss in [NSZ16] can enable better target localization with less obscurity.

IV. Measurement setup for concept verification

All this while, most MIMO radar research has been focused on theoretical findings and simulations verifications. The ultimate reason is because of the high cost for a dedicated equipment as well as the lack of commercially available platforms. Although Software Defined Radios are prime candidates for the testing of communication systems, their limited analog bandwidth meant that they were unsuitable for radars. Moreover, for each additional transmitter or receiver, an additional RF frontend and analog-to-digital or digital-to-analog converters (ADC, DAC) will be needed.

Most researchers hence experiment with time division multiplexed (TDM) schemes such as Pasya et al. in [PIK14], where multiple measurements are taken via antennas that are adjusted with stepped motors to emulate an antenna array. While this mode of measurement verification is sufficient, it does not take much hardware effect into account i.e. antenna coupling, synchronization between multiple transmitters and receivers, etc. Above all, measurements with such a setup cannot be conducted outdoors especially with moving targets, since the scenario changes too fast during the time required to shift the antennas.

A Universal Software Radio Peripheral (USRP) model X310 capable of an instantaneous analog bandwidth of 120 MHz only became available in 2014. The measurement verifications of the 3D+velocity capability as well as the interference cancellation algorithm are performed using a 4×4 MIMO setup[5] using four of the X310 USRPs. The operating frequency is scaled to 4.05 GHz due to the upper limit of the USRPs but the instantaneous signal bandwidth used is nevertheless a wideband of 100 MHz.

At the time of writing, this is the first reported real-time setup using a commercially available hardware, with velocity measurements. Real-time indoor electromagnetic compatibility chamber (EMC) and outdoor measurements are made and discussed in Chapter 7. For the indoor measurements static targets (corner reflectors) as well as a Doppler emulator are used to verify the RadCom concepts presented in the earlier chapters. In the outdoor measurements, velocity measurements were made with a van traveling at around 30 km/h. All measurement results correspond well with their expected outcome.

Chapter 8 then wraps up the work presented in this thesis and highlights all major contributions.

5 i.e. four transmitters and four receivers

2 Signal model for multi-antenna RadCom network systems

This chapter begins with the fundamentals that make up the OFDM-based radar. The features of OFDM and its use in a SISO radar setup are first presented, along with the radar quality parameters i.e. the maximum unambiguous range and Doppler, and their resolutions. Next, the SISO signal model is expanded to a MIMO one to accommodate a multi-antenna radar node. The effect of this signal model change on the radar's quality parameters will be discussed in detail along with the method for range and Doppler estimations.

2.1 OFDM signals

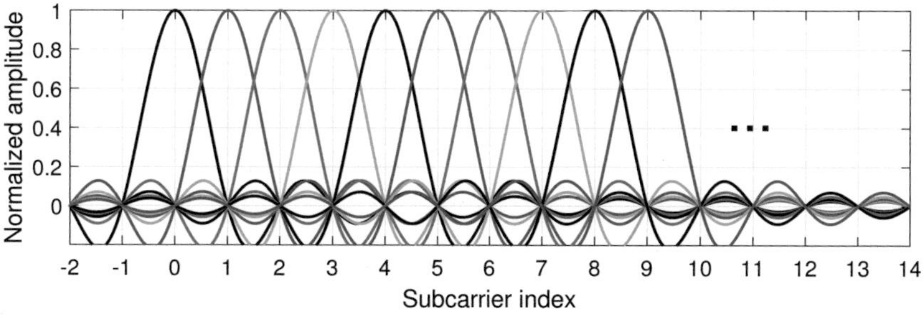

Fig. 2.1: OFDM subcarriers visualization.

The OFDM signals can be visualized as a set of signals spanning a bandwidth, all modulated onto non-overlapping subcarrier frequencies to create multiple subchannels and then transmitted in parallel in time. If the bandwidth of each subcarrier is sufficiently small to allow a large number of parallel channels to span the above-said signal bandwidth, a long bit stream can be divided among these subchannels. In doing so the rate of transfer can be decreased yet the parallel streaming helps achieve the same throughput rate.

In order to avoid interference between channels, the subcarrier frequencies are chosen in such a way that they are orthogonal to each other. Let Δf be the subcarrier spacing (the space between the peaks of the sinc function as shown in Fig. 2.1), T_0 the symbol duration, n the subcarrier index and N the number of subcarriers within the signal bandwidth. Choosing Δf to be multiples of T_0 will eliminate symbol and subcarrier overlapping. In order for the orthogonality to span over all N subcarriers, the setting $n\Delta f = n/T_0$ for $n = 0, 1, ..., N-1$ is used. All subcarriers are now regularly spaced in frequency domain, so that they transmit at integer multiples of the first carrier frequency as depicted in Fig. 2.1. It can be seen that at every sinc signal peak, there are zero-crossings from all other neighboring sinc signals. Hence one T_0 contains N orthogonal symbols and this is known as one OFDM symbol. This explains how the same throughput as a high-rate single frequency carrier signal can be achieved. Also, since the individual bandwidth of the subcarriers are sufficiently small, they are usually much smaller than the channel's coherent bandwidth, therefore each subchannel experiences flat-fading. This is the notion behind the 'robustness' of OFDM signals against frequency selectivity in channels.

2.1.1 OFDM signal generation

To represent a bit stream, let there be a matrix of the length N of modulated arbitrary symbols to represent one OFDM symbol. The long bit stream will span over a series of M subsequent OFDM symbols with $m = 0, 1, ..., M-1$ being the OFDM symbol index to result in the matrix $\mathbf{A}(n,m)$. At the m-th OFDM symbol, the matrix can be written as

$$\mathbf{A}_m(n) = \begin{bmatrix} a_{0,m} & a_{1,m} & a_{2,m} & \cdots & a_{N-1,m} \end{bmatrix}^{\mathsf{T}}, \tag{2.1}$$

with $[\cdot]^{\mathsf{T}}$ indicating a transpose matrix. It is clear here that the bit stream consists of modulated symbols $a_{n,m}$ generated in the frequency domain and is thus in discrete signal form. To convert this signal to its analog equivalent for transmission, an inverse Fourier transform is required. In literature, the baseband signal of the m-th symbol is frequently written as

$$x_m(t) = \sum_{n=0}^{N-1} a_m(n)\,\mathrm{e}^{\mathrm{j}2\pi n\Delta f t} = N \cdot \mathcal{F}^{-1}\left[\mathbf{A}_m\right], \tag{2.2}$$

with $\mathcal{F}^{-1}\{\cdot\}$ denoting the inverse Fourier transform. The exponential term after the modulation symbol represents a phase rotation in time domain due to the frequency shift Δf in the frequency domain as constituted by the Fourier transform pair of (valid for continuous and discrete signal forms), given by

$$x(t) \cdot \mathrm{e}^{\mathrm{j}2\pi f_0 t} \quad \bullet\!\!-\!\!\circ \quad X(f - f_0). \tag{2.3}$$

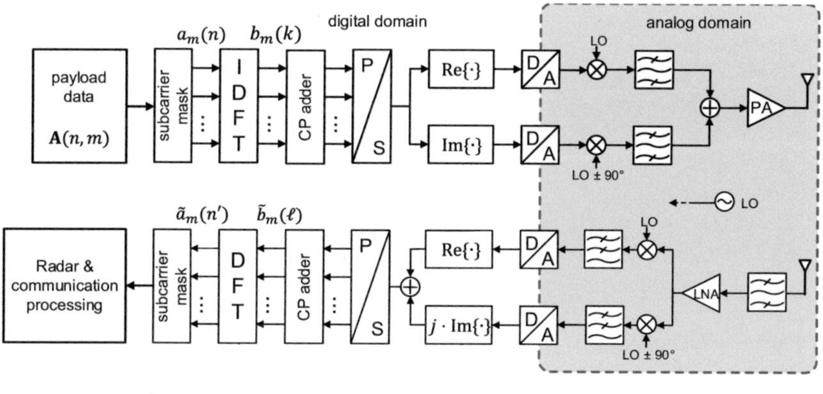

Fig. 2.2: SISO OFDM radar and communication block diagram.

This means that the modulation symbol in frequency domain is phase rotated according to the subcarrier frequency in time domain. The domain before the digital-to-analog (DA) conversion and up-conversion to the operating frequency is actually the discrete time domain as shown in the simplified SISO radar and communication block diagram in Fig. 2.2, which consist of the transmit and receive chains. The baseband modulated symbols will first pass through a subcarrier mask, which determines the subcarriers that are enabled or disabled as sometimes subcarriers around the center and edges of the spectrum are disabled to avoid the local oscillator (LO) carrier leakage and filter roll-off effect respectively.

Then the modulation symbols are inverse Fourier transformed to a discrete time signal where each OFDM symbol is then prepended with a cyclic prefix then converted to a serial signal. Since this signal output is still in the complex domain, their real (Re{·}) and imaginary (Im{·}) components are extracted to make both real-numbered signal streams. After the parallel DA conversion, both signal streams are now analog continuous signals. The stream containing the real components are up-converted by the LO to result in the radio frequency (RF) In-Phase (I) signal, while the stream with the previously imaginary components are up-converted with a 90° phase offset to yield the Quadrature (Q) signal. The I&Q modulator is used to eliminate one of the resulting sidebands. Both I and Q streams are then bandpass filtered before being added to be amplified by a power amplifier (PA) then transmitted over antennas. Hence the generation of the transmit signal is more accurately expressed as

$$\begin{aligned} x_m(k) = b_m(k) &= \sum_{n=0}^{N-1} a_m(n)\,\mathrm{e}^{\mathrm{j}2\pi n \Delta f k T_\mathrm{S}}, \quad T_\mathrm{S} = \tfrac{T_0}{N} = \tfrac{1}{N\Delta f} \\ &= \sum_{n=0}^{N-1} a_m(n)\,\mathrm{e}^{\mathrm{j}\frac{2\pi}{N}nk} \\ &= N \cdot \mathrm{IDFT}\left[\mathbf{A}_m\right], \end{aligned} \tag{2.4}$$

where T_S is the sampling time and IDFT denotes the inverse discrete Fourier transform, often used in mathematical softwares. It becomes clear here that the OFDM symbols are generated digitally as already 'sampled' signals therefore the sampling frequency $f_\mathrm{S} = 1/T_\mathrm{S}$ for the OFDM signal is also equal to its bandwidth (in the case where all subcarriers are used).

At the receiver (Rx) side, the RF signal is first amplified by a low-noise amplifier (LNA) then split into two streams to be down-converted simultaneously by an in-phase and quadrature mixer (LO) as used in the transmit chain. These two streams of down-converted analog signals are then put through a lowpass filter to eliminate the higher frequencies then converted by the analog-to-digital converter (ADC) to revert them back into discrete time signals. Assuming that the Nyquist sampling has been fulfilled with no time offset i.e. correct sampling at the peaks of the time-limited down-converted RF signal, the modulation symbols are recovered through the inverse of the process at the transmit chain.

The signal is put through the discrete Fourier transform (DFT) block to yield the recovered modulation symbols

$$a_m(n) = \sum_{n=0}^{N-1} x_m(k)\,\mathrm{e}^{-\mathrm{j}\frac{2\pi}{N}nk} = \mathrm{DFT}\left[x_m\right]. \tag{2.5}$$

2.1.2 Up-conversion to a carrier frequency:

The spectrum of the OFDM signal, whether in baseband or RF level, is always situated with the carrier frequency in the middle. The subcarrier at the center of the spectrum is normally denoted with the index 0 making the subcarrier indices for N subcarriers $n = \{-\frac{N}{2}, -\frac{N}{2}+1, ..., -1, 0, 1, ..., \frac{N}{2}-1\}$. To simplify the notation however, the subcarrier index according to the notation $n = \{0, 1, ..., N-1\}$ is used throughout this thesis and the subcarrier index $n = \frac{N}{2}+1$ is hence in the middle. This is because the generation of the modulation symbol in mathematical software cannot be indexed with negative numbers.

2.1.3 Cyclic prefix

A cyclic prefix (CP) is basically a copy of the tail of an OFDM time-domain signal that is prepended at the beginning of the signal to create a guard interval (GI). The guard interval adds immunity against Inter-symbol Interference (ISI) while the cyclic nature of this interval (instead of a blank guard interval) ensures a correct sampling interval at the receiver. Adding a CP also turns the linear convolution into a circular convolution in time domain, which then translates to a multiplication with the DFT operation and simplifies the signal processing at the receiver considerably. The frequency domain channel transfer function can then be obtained using a simple element-wise division operation of the received and transmitted symbols as will be shown in Section 2.5 and Chapter 5. Denoting the duration of N_{CP} samples of the cyclic prefix as T_{CP}, the number of discrete time samples per OFDM symbol is then $N + N_{\mathrm{CP}}$ and the total duration of one OFDM symbol is now lengthened to $T = T_0 + T_{\mathrm{CP}}$.

2.1.4 Nomenclature and notation

Before proceeding to the fundamentals of the OFDM radar, a briefing of the nomenclature and notations that will be used throughout this thesis is in order. The notation for a matrix is always in bold font, for example $\mathbf{A}(n,m)$ or $\mathbf{A}$. The size of the matrix is denoted as $\mathbb{C}^{N\times M}$ where N is the number of rows and M the number of columns in the matrix. To represent the n-th row and m-th column element of the $\mathbf{A}(n,m)$ matrix, the notation $A_{n,m}$ or the specified element symbol within the matrix e.g. $a_{n,m}$ is used. Vectors, will always have an arrow sign above the letters e.g. $\vec{a}$.

The '**modulation symbol**' is a baseband data that is modulated with arbitrary modulation schemes such as the PSK[6] and QAM[7]. The data can be of text, graphical or sound origins as is commonly called the '**payload data**'. The range of the values of the modulation symbols in terms of their amplitude and phase is contained in the modulation alphabet, which is denoted by $\mathcal{A}$. The OFDM signal is generated in the frequency domain with N modulation symbols making one '**OFDM symbol**'. A subsequent series of M transmit OFDM symbols makes one '**frame**'. The rows of the OFDM frame constitutes the subcarrier index of $n = 0,1,2,...,N-1$ and the columns represent the symbol index of $m = 0,1,2,...M-1$. Hence the row axis of the OFDM frame is also called the '**frequency-subcarrier axis**' while the column axis is the '**time-symbol axis**', to differentiate them from the general frequency and time axes. The abbreviations '**Tx**' and '**Rx**' will be used to refer to the transmitter and the receiver directly.

At the baseband level one OFDM frame comprising the modulation symbols to be transmitted by one transmit antenna can be written as

$$\mathbf{A}(n,m) = \begin{pmatrix} a_{0,0} & a_{0,1} & \cdots & a_{0,M-1} \\ a_{1,0} & a_{1,1} & \cdots & a_{1,M-1} \\ \vdots & \vdots & \ddots & \vdots \\ a_{N-1,0} & a_{N-1,1} & \cdots & a_{N-1,M-1} \end{pmatrix} \in \mathcal{A}^{N\times M}. \tag{2.6}$$

6 PSK - Phased Shift Keying, particularly quadrature PSK or QPSK
7 QAM - Quadrature Amplitude Modulation

2.2 Fundamentals of SISO OFDM radar

To start with using OFDM signals as radar signals, let us consider a SISO quasi-monostatic radar here, whereby there is only one Tx and Rx pair on the same platform, denoted by T_p and R_q in Fig. 2.3(a). The Tx and Rx are said to be 'collocated' because their distance to each other d is sufficiently smaller than their distance to the target r_h, in their antenna far-field region. The radar transmits a signal $x(t)$ in time domain and it can be represented as

$$x(t) = \sum_{m=0}^{M-1} \sum_{n=0}^{N-1} A_{n,m} \mathrm{e}^{\mathrm{j}2\pi f_n t} \mathrm{rect}\left(\frac{t - mT_0}{T_0}\right). \tag{2.7}$$

This simply means that each modulation symbol $A_{n,m}$ at the n-th row and m-th column of the data matrix $\mathbf{A}(n,m)$ is modulated onto the subcarrier with the index n according to $f_n = n\Delta f$. The function $\mathrm{rect}\left(\frac{t-mT_0}{T_0}\right)$ is a rectangular function with the duration of mT_0 governing the length of every m-th OFDM symbol in time domain. This function equivalent to a sinc function in the frequency domain, which is a more common representation of one orthogonal subcarrier with the subcarrier spacing of $T_0 = 1/\Delta f$. The CP is not shown for simplicity since it will be removed before the radar processing and will not affect the outcome of the radar estimation when certain conditions are met, as will be seen in Section 2.5.6.

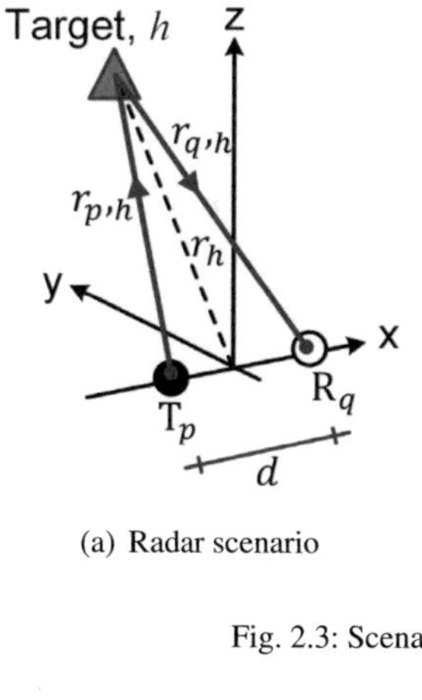

(a) Radar scenario

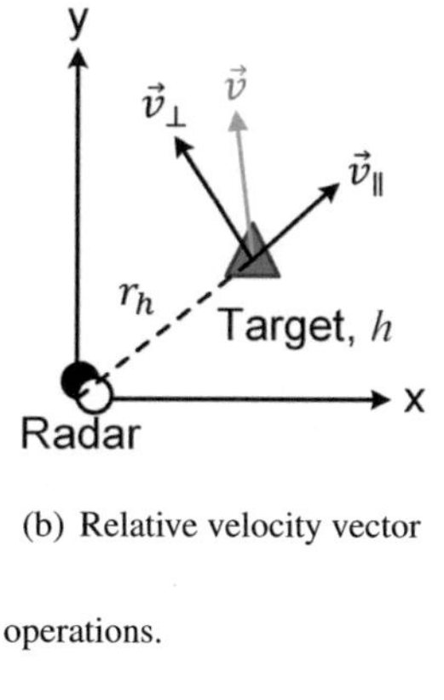

(b) Relative velocity vector

Fig. 2.3: Scenario of radar operations.

Moving on to the scenario for radar estimation, also called the 'radar channel'; let there be an arbitrary number of H (with the index $h = 0, 1, ..., H-1$) targets or objects, each located at the distance r_h away from the radar as depicted in Fig. 2.3(a). The incident wave of $x(t)$ hits the objects at the euclidean distance of $r_{p,h}$ and is then backscattered to the Rx with the distance $r_{q,h}$. Since the Tx and Rx are collocated, meaning $d \ll r_h$, it can be assumed that the total distance traveled is $r_{p,h} + r_{q,h} \approx 2r_h$. Accordingly a phase term of $\mathrm{e}^{-\mathrm{j}2\pi f_n 2r_h/c_0}$ will be incurred to $x(t)$ since a time delay becomes a phase rotation when Fourier transformed to the frequency domain according to

$$x(t-\tau) \quad \bullet\!\!-\!\!\circ \quad X(f) \cdot \mathrm{e}^{-\mathrm{j}2\pi f \tau}. \tag{2.8}$$

If the objects are moving, a Doppler shift due to the objects' radial velocities will be incurred. As shown in Fig. 2.3(b), the target's real velocity $\vec{v}$ can be decomposed to the velocity that is parallel to the target's euclidean distance vector $\vec{v}_{\|}$, and the velocity that is perpendicular to the distance vector $\vec{v}_{\perp}$, making $\vec{v} = \vec{v}_{\|} + \vec{v}_{\perp}$. At the radar only the term equivalent to $2\vec{v}_{\|}$ (also written as the relative velocity $2v_{\mathrm{rel}}$) due to the wave traveling to and from the target, can be 'seen' and hence estimated.

Considering the above-mentioned scenario, the receive analog signal in time domain before down-conversion from the carrier frequency f_c has the form of

$$y(t) = \sum_{h=0}^{H-1} \alpha_h \mathrm{e}^{\mathrm{j}\zeta} \sum_{m=0}^{M-1} \sum_{n=0}^{N-1} a_{n,m} \left[\mathrm{e}^{\mathrm{j}2\pi(f_n+f_c)\left(t-2\left(\frac{r_h - v_{\mathrm{rel},h} t}{c_0}\right)\right)} \right] \cdot \mathrm{rect}\left(\frac{t - mT_0 - \tau_{h,\max}}{T_0 + \tau_{h,\max}} \right) + z(t), \tag{2.9}$$

with τ_h being the duration taken to traverse $2r_h$, $\tau_{h,\max}$ is the round-trip time of the of the farthest target and $\mathrm{e}^{\mathrm{j}\zeta}$ takes all random phase rotations into account. The OFDM symbol duration is also extended by $\tau_{h,\max}$, as given in the term $\mathrm{rect}\left(\frac{t-mT_0-\tau_{h,\max}}{T_0+\tau_{h,\max}}\right)$. Expanding Eq. (2.9) leads to

$$\begin{aligned}
y(t) &= \sum_{h=0}^{H-1} \alpha_h \mathrm{e}^{\mathrm{j}\zeta} \sum_{m=0}^{M-1} \sum_{n=0}^{N-1} a_{n,m} \left[\mathrm{e}^{\mathrm{j}2\pi f_n t} \mathrm{e}^{-\mathrm{j}2\pi f_n \frac{2r_h}{c_0}} \mathrm{e}^{-\mathrm{j}2\pi f_c \frac{2r_h}{c_0}} \mathrm{e}^{\mathrm{j}2\pi f_c t} \mathrm{e}^{\mathrm{j}2\pi (f_n + f_c) \frac{2v_{\mathrm{rel},h}}{c_0} t} \right] \cdot \\
&\qquad \mathrm{rect}\left(\frac{t - mT_0 - \tau_{h,\max}}{T_0 + \tau_{h,\max}} \right) + z(t) \\
&\approx \sum_{h=0}^{H-1} \alpha_h \mathrm{e}^{\mathrm{j}\zeta} \sum_{m=0}^{M-1} \sum_{n=0}^{N-1} a_{n,m} \left[\mathrm{e}^{\mathrm{j}2\pi f_n t} \mathrm{e}^{-\mathrm{j}2\pi f_n \tau_h} \mathrm{e}^{-\mathrm{j}2\pi \frac{2r_h}{\lambda_c}} \mathrm{e}^{\mathrm{j}2\pi f_c t} \mathrm{e}^{\mathrm{j}2\pi \frac{2v_{\mathrm{rel},h}}{\lambda_c} t} \right] \cdot \\
&\qquad \mathrm{rect}\left(\frac{t - mT_0 - \tau_{h,\max}}{T_0 + \tau_{h,\max}} \right) + z(t), \qquad f_c \gg f_n.
\end{aligned} \tag{2.10}$$

Since the carrier frequency is normally in the gigahertz range, which is much larger than the baseband subcarrier spacing (in kilohertz range), it is assumed that the Doppler term affects all subcarriers by an equal amount as can be seen in the term $\mathrm{e}^{\mathrm{j}2\pi (f_n + f_c) \frac{2v_{\mathrm{rel},h}}{c_0} t} \approx \mathrm{e}^{\mathrm{j}2\pi \frac{2v_{\mathrm{rel},h}}{\lambda_c} t} = \mathrm{e}^{\mathrm{j}2\pi f_{\mathrm{D},h} t}$ in Eq. (2.10). After down-conversion to the baseband level, the term $\mathrm{e}^{\mathrm{j}2\pi f_c t}$ vanishes leaving

$$\begin{aligned}
y(t) &= \sum_{m=0}^{M-1} \sum_{n=0}^{N-1} a_{n,m} \mathrm{e}^{\mathrm{j}2\pi f_n t} \left[\sum_{h=0}^{H-1} \alpha_h \mathrm{e}^{\mathrm{j}\zeta} \mathrm{e}^{-\mathrm{j}2\pi f_n \tau_h} \mathrm{e}^{-\mathrm{j}2\pi \frac{2r_h}{\lambda_c}} \mathrm{e}^{\mathrm{j}2\pi \frac{2v_{\mathrm{rel},h}}{\lambda_c} t} \right] \cdot \\
&\qquad \mathrm{rect}\left(\frac{t - mT_0 - \tau_{h,\max}}{T_0 + \tau_{h,\max}} \right) + z(t) \\
&= \sum_{m=0}^{M-1} \sum_{n=0}^{N-1} a_{n,m} \mathrm{e}^{\mathrm{j}2\pi f_n t} \left[\sum_{h=0}^{H-1} \alpha_h \mathrm{e}^{\mathrm{j}\zeta} \mathrm{e}^{-\mathrm{j}2\pi f_n \tau_h} \mathrm{e}^{\mathrm{j}\Psi_h} \mathrm{e}^{\mathrm{j}2\pi f_{\mathrm{D},h} t} \right] \cdot \\
&\qquad \mathrm{rect}\left(\frac{t - mT_0 - \tau_{h,\max}}{T_0 + \tau_{h,\max}} \right) + z(t) \\
&= \sum_{m=0}^{M-1} \sum_{n=0}^{N-1} \hat{a}_{n,m} \mathrm{e}^{\mathrm{j}2\pi f_n t} + z(t).
\end{aligned} \tag{2.11}$$

Comparing Eq. (2.7) with Eq. (2.11), it can be see that $\hat{a}_{n,m}$ is the received modulation symbol, which contains the original transmit symbol $a_{n,m}$ and all the associated channel effects. The channel effects in Eq. (2.11) can be decomposed to six effects as follows:

- α_h is the attenuation factor of the reflected signal derived from the monostatic radar equation, where

$$\begin{aligned} \frac{P_\mathrm{R}}{P_\mathrm{T}} &= \frac{\lambda_\mathrm{c}^2 \sigma_h}{(4\pi)^3 r_h^4} G_\mathrm{T} G_\mathrm{R} \\ \alpha_h &= \sqrt{\frac{\lambda_\mathrm{c}^2 \sigma_h}{(4\pi)^3 r_h^4} G_\mathrm{T} G_\mathrm{R}}. \end{aligned} \tag{2.12}$$

  P_T and P_R are the transmit and receive power respectively, G_T and G_R refers to the Tx and Rx antenna gains respectively, $\lambda_\mathrm{c} = c_0 / f_\mathrm{c}$ is the operating wavelength with c_0 being the speed of light and f_c the operating frequency, and σ_h is the h-th object's radar cross section (RCS). It is assumed that the attenuation factor affects all modulation symbols of the same Tx signal equally (slow fading channel).

- The term $e^{\mathrm{j}\Psi_h}$ is the phase offset due to the euclidean distance of the objects to the radar. It is constant over all modulation symbols of the same frame and will not interferer with the estimation of the range and Doppler terms. With a multi-antenna radar configuration that can exploit spatial diversity however, this term is essential to estimate the position of the objects in the azimuth and/or the elevation, as will be discussed in Chapter 3.

$$\Psi_h = -2\pi \frac{|2r_h|}{\lambda_\mathrm{c}} \tag{2.13}$$

- The term $\mathrm{e}^{-\mathrm{j}2\pi f_n \tau_h}$ is related to the round-trip time delay of the signal in terms of a phase rotation through the frequency-subcarrier axis with

$$\tau_h = \frac{2r_h}{c_0}. \tag{2.14}$$

- The $\mathrm{e}^{\mathrm{j}2\pi f_{\mathrm{D},h} t}$ term is the Doppler shift term. Just like the distance term, the signal experiences two times the Doppler shifts while traveling to and back from the objects, whereby

$$f_{\mathrm{D},h} = \frac{2 v_{\mathrm{rel},h}}{\lambda_\mathrm{c}}. \tag{2.15}$$

- $e^{\mathrm{j}\zeta}$ accounts for any random phase offset due to the state of the channel or hardware and is hence a complex value.

- The final term is the white Gaussian noise term $z(t)$.

The evaluation of the range and Doppler (or relative velocity) of the targets can be done by comparing the transmit and receive signals to extract the differences, which in this case are the phase rotation terms over the frequency-subcarrier and time-symbol axes. While the conventional cross-correlation method with Doppler filter banks can be used, a more efficient method that requires less computational effort is used in this work and is presented in Section 2.5.

2.3 Radar quality parameters

For any SISO radar, there are four quality parameters that reflects the capability of the radar's estimation. They are as follows:

Maximum unambiguous range, r_{max} refers to the theoretical maximum possible target distance detectable by the radar without any ambiguity. Radars can only estimate the change of phase due to the range over $[0, 2\pi]$ or $[-\pi, \pi]$. Within this 'wrapped phase', each point of the time delay τ_h will incur a different phase term. Range ambiguity arises when the 2π limit is exceeded, causing the target to appear at distance, which is the vector norm of the $\mathrm{e}^{-\mathrm{j}2\pi f_n \tau_h}$ term. For example, a target detected at a distance, which incurs the phase of $\mathrm{e}^{\mathrm{j}3\pi}$ will appear at the distance corresponding to $\mathrm{e}^{\mathrm{j}\pi}$. In reality however, the furthest object that can be detected by the radar is dependent mainly on the radar's transmit power, the object's RCS (the power backscattered from the object to the radar), the terrain (multipaths, clutter, etc), and receiver sensitivity. For OFDM signals r_{max} is given by

$$r_{\text{max}} = \frac{c_0 T_0}{2} = \frac{c_0}{2\Delta f}. \tag{2.16}$$

Range resolution, Δr can be viewed in such a way, that when there are two objects at Δr apart, then they can be distinguished as two separate objects. When they are less than Δr apart, they will appear as the same point on the radar image or their peaks will not be far enough apart to determine that they are from two separate objects. Hence the finer the range resolution, the better the separability of the objects. The range resolution is dependent on the signal bandwidth BW and is given by

$$\Delta r = \frac{c_0}{2N\Delta f} = \frac{c_0}{2BW}. \tag{2.17}$$

Maximum unambiguous Doppler, $f_{\mathrm{D,max}}$ is analogous to r_{max} and along with the maximum unambiguous velocity, is given by

$$\begin{aligned} f_{\mathrm{D,max}} &= \frac{1}{T} = \frac{2v_{\mathrm{max}}}{\lambda_{\mathrm{c}}} \\ \Rightarrow v_{\mathrm{max}} &= \frac{\lambda_{\mathrm{c}}}{2T} = \frac{c_0}{2Tf_{\mathrm{c}}}. \end{aligned} \tag{2.18}$$

In terms of OFDM subcarrier spacing, the term $f_{\mathrm{D,max}}$ can affect the subcarrier in the positive and negative directions from the center point of the subcarrier. Hence the actual maximum unambiguous Doppler shift and velocity must be divided into two, representing the two halves of the OFDM subcarrier from its center to yield

$$\begin{aligned} f_{\mathrm{D,max}} &= \pm\frac{1}{2T} = \frac{2v_{\mathrm{max}}}{\lambda_{\mathrm{c}}} \\ \Rightarrow v_{\mathrm{max}} &= \pm\frac{\lambda_{\mathrm{c}}}{4T} = \pm\frac{c_0}{4Tf_{\mathrm{c}}}. \end{aligned} \tag{2.19}$$

As can be seen, the maximum unambiguous velocity is also dependent on the total duration of the OFDM transmit symbol hence it is also influenced by the duration of the CP. The longer the CP length, the less the maximum unambiguous Doppler detectable by the radar as given by

$$f_{\mathrm{D,max}} = \pm\frac{1}{2}\left(\frac{1}{T_0 + T_{\mathrm{CP}}}\right). \tag{2.20}$$

Doppler resolution, Δf_{D} is analogous to Δr. Along with the corresponding velocity resolution, it is given by

$$\begin{aligned} \Delta f_{\mathrm{D}} &= \frac{1}{MT} \\ \Delta v &= \frac{\lambda_{\mathrm{c}}}{2MT} = \frac{c_0}{2MTf_{\mathrm{c}}}. \end{aligned} \tag{2.21}$$

2.4 Signal model for a multi-antenna OFDM-based radar system

Now that the SISO OFDM radar fundamentals have been introduced, the signal mode will be extended to cater to a multi-antenna or MIMO configuration. A notation at the baseband modulation symbol level is also used from this point onward instead of the time-domain expressions to better represent the signal processing steps. Here, the radar transmit signal, channel and receive signal for one MIMO OFDM radar node will be analyzed in detail, extending the works on the SISO OFDM radar done by Sturm [Stu12] and Braun [Bra14].

In order to accommodate multiple-antennas (which means, multiple parallel transmit channels) without intra-system interference the bandwidth of the signal is divided into '**channels**' or '**users**'. In the simplest form, this means that the N available subcarriers are assigned to the multiple-antennas based on their allocated '**user index**' to ensure that every Tx antenna will transmit on one non-overlapping '**subcarrier set**'. One of the strategies for subcarrier assignment is presented in detail in [SSBZ13] and [Stu12]. Although this work was made for multiple-user access of SISO OFDM radar systems, this strategy is however also valid for this multi-antenna OFDM radar. The subcarrier assignment strategy is described as follows.

Let there be N_{ch} channels within the signal bandwidth. Each non-overlapping channel can be assigned with a user index of $u = 0, 1, ..., N_{\text{ch}} - 1$ with each user allocated $\frac{N}{N_{\text{ch}}}$ number of subcarriers. The relation of user index to its set of allocated subcarriers is given by

$$\begin{aligned} n_u &= u + iN_{\text{ch}}, \qquad u = 0, 1, ..., N_{\text{ch}} - 1, \;\; i = 0, 1, ..., \frac{N}{N_{\text{ch}}} - 1, \\ \therefore f_{n_u} &= n_u \Delta f. \end{aligned} \tag{2.22}$$

This shows that the u-th user will transmit only on every N_{ch}-th subcarrier and this subcarrier set denoted by n_u will span the entire signal bandwidth in an interleaved manner, hence the name 'spectrally interleaved OFDM signals' used in [SSBZ13]. It can be seen that the counter i is a sample of the subcarrier index n at every N_{ch} interval with the condition $0 \leq iN_{\text{ch}} \leq N - 1$.

An example with $N_{\text{ch}} = 4$ is shown in Fig. 2.1, whereby each of the same colored OFDM subcarrier belongs to the set of subcarriers allocated to the same u. The block diagram of the MIMO OFDM radar's multiple transmit and receive chains is as shown in Fig. 2.4. It can be seen that the differences compared to the SISO block in Fig. 2.2 are the multiple streams in the digital domain and RF frontend (denoted by the analog domain). For every additional antenna added to the radar node, one separate transmit or receive chain will be required. The subcarrier mask now functions also to filter the subcarrier sets for use as assigned to the particular transmit chain, besides the initial aforementioned functions.

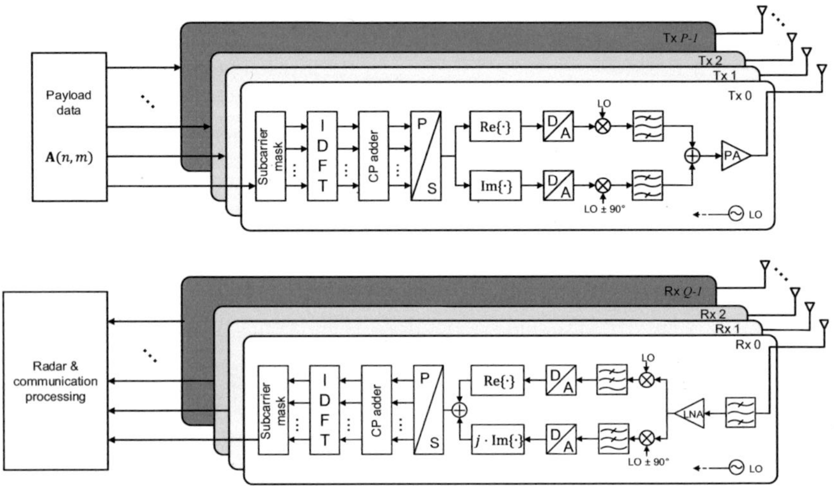

Fig. 2.4: MIMO OFDM radar and communication block diagram. Each antenna requires its own transmit or receive chain i.e. RF frontend and digital stream.

This spectrally interleaved OFDM signal model has been chosen instead of other subcarrier allocation strategies to take into account the signal model's effect on the radar's quality parameters. By spanning each subcarrier set over the whole signal bandwidth, the range resolution can be retained. The maximum unambiguous range of the radar however will drop by a factor of N_{ch} according to the expression

$$r_{\mathrm{max,interleaved}} = \frac{c_0}{2N_{\mathrm{ch}}\Delta f}. \tag{2.23}$$

Intuitively this can be seen as a widening of the subcarrier spacing by a factor of N_{ch}. This signal model however imposes no effect on the radar's Doppler estimation properties.

2.4.1 Transmit signal

Let one OFDM node be consist of P transmit- and Q receive antennas. This configuration is commonly denoted as $P \times Q$. Let the Tx and Rx antenna index be $p = 0, 1, ..., P-1$ and $q = 0, , ..., Q-1$ respectively. Now, the data to be transmitted from the p-th Tx must be modulated onto the correct set of assigned subcarriers. For illustration, consider there to be two channels and a radar node with two Tx antennas, where $N_{\text{ch}} = P = 2$. Hence the transmit matrix for each antenna can be represented as $\mathbf{A}_u = \mathbf{K}_u^{\text{mask}} \circ \mathbf{A}$, where $\mathbf{K}_u^{\text{mask}}$ is the subcarrier mask for the user index u. The symbol $[\circ]$ represents the Hadamard product (or element-wise multiplication) for matrices. If the Tx $p = 0$ and $p = 1$ are assigned the user index $u = 0$ and $u = 1$ respectively, the mapping of the modulation symbols of both the Tx are written as

$$\mathbf{A}_1(n,m) = \begin{pmatrix} 1 & 1 & \cdots & 1 \\ 0 & 0 & \cdots & 0 \\ 1 & 1 & \cdots & 1 \\ 0 & 0 & \cdots & 0 \\ \vdots & \vdots & \ddots & \vdots \\ 1 & 1 & \cdots & 1 \\ 0 & 0 & \cdots & 0 \end{pmatrix} \circ \mathbf{A} = \begin{pmatrix} a_{0,0} & a_{0,1} & \cdots & a_{0,M-1} \\ 0 & 0 & \cdots & 0 \\ a_{2,0} & a_{2,1} & \cdots & a_{2,M-1} \\ 0 & 0 & \cdots & 0 \\ \vdots & \vdots & \ddots & \vdots \\ a_{N-2,0} & a_{N-2,1} & \cdots & a_{N-2,M-1} \\ 0 & 0 & \cdots & 0 \end{pmatrix}$$

$$\mathbf{A}_2(n,m) = \begin{pmatrix} 0 & 0 & \cdots & 0 \\ 1 & 1 & \cdots & 1 \\ 0 & 0 & \cdots & 0 \\ 1 & 1 & \cdots & 1 \\ \vdots & \vdots & \ddots & \vdots \\ 0 & 0 & \cdots & 0 \\ 1 & 1 & \cdots & 1 \end{pmatrix} \circ \mathbf{A} = \begin{pmatrix} 0 & 0 & \cdots & 0 \\ a_{1,0} & a_{1,1} & \cdots & a_{1,M-1} \\ 0 & 0 & \cdots & 0 \\ a_{3,0} & a_{3,1} & \cdots & a_{3,M-1} \\ \vdots & \vdots & \ddots & \vdots \\ 0 & 0 & \cdots & 0 \\ a_{N-1,0} & a_{N-1,1} & \cdots & a_{N-1,M-1} \end{pmatrix}. \quad (2.24)$$

The first Tx will transmit on $n_{u=p=0} = \{0, 2, 4, ..., N-2\}$, which is on all even subcarrier indices and the second Tx on $n_{u=p=1} = \{1, 3, 5, ..., N-1\}$, on all odd subcarrier indices. In the matrices in Eq. (2.24) the row with a '0' is means that there is no payload data. Since the user index u can be mapped arbitrarily to the antenna p, the transmit matrix is connected to the modulation symbol matrix by

$$\mathbf{X}_p(n_u, m) = \mathbf{A}_u(n_u, m). \tag{2.25}$$

The other parameters for the spectrally interleaved OFDM signal are the same as for the SISO case and are summarized in Table 2.1.

Table 2.1: OFDM RadCom parameter symbols.

Symbol	Parameter
$\mathcal{A}$	Modulation alphabet
N	Number of subcarriers within the bandwidth with $n = 0, 1, ..., N-1$
M	Number of contiguous OFDM symbols with $m = 0, 1, ..., M-1$
N_{CP}	Number of guard interval/cyclic prefix symbols
Δf	Subcarrier spacing
$T_0 = 1/\Delta f$	OFDM symbol duration
T_{CP}	Duration of guard interval/ cyclic prefix
$T = T_0 + T_{\text{CP}}$	Total duration of one OFDM symbol
$BW = N \cdot \Delta f$	Total signal bandwidth
f_{c}	Carrier frequency
$\lambda_{\text{c}} = c_0 / f_{\text{c}}$	Wavelength of f_{c}
P	Number of transmit antennas with $p = 0, 1, ..., P-1$
Q	Number of receive antennas with $q = 0, 1, ..., Q-1$
N_{ch}	Number of user channels with $1 \leq N_{\text{ch}} \leq N$
u	User index with $u = 0, 1, ..., N_{\text{ch}} - 1$
i	Counter for subcarrier indices associated with u, where $i = 0, 1, ..., \frac{N}{N_{\text{ch}}} - 1$
$n_u = u + iN_{\text{ch}}$	Set of subcarrier indices allocated to u

2.4.2 Radar channel

Here the radar channel is defined for the h-th target out of H number of arbitrary targets in the scenario or radar channel in terms of time delay (range) and Doppler (relative velocity), building upon the basics introduced in Section 2.2.

Effect of target's range

As can be seen from Eq. (2.11), the range phase rotation term is dependent on the subcarrier index indicated by f_n. With the interleaved OFDM signal model not all subcarriers are used by the p-th Tx hence a new description of the range phase rotation $\kappa_{r_h}(n_u)$ over the assigned subcarrier set of n_u is in order.

Neglecting the Tx and Rx antenna indices and taking only the user index into account, the range phase rotation can be expanded as

$$\begin{aligned}
\kappa_{r_h}(n_u) &= \mathrm{e}^{-\mathrm{j}2\pi f_{n_u}\tau_h} \\
&= \mathrm{e}^{-\mathrm{j}2\pi(u+iN_{\mathrm{ch}})\Delta f\tau_h} \\
&= \mathrm{e}^{-\mathrm{j}2\pi u\Delta f\tau_h}\cdot \mathrm{e}^{-\mathrm{j}2\pi(iN_{\mathrm{ch}})\Delta f\tau_h}, \qquad i=0,1,...,\frac{N}{N_{\mathrm{ch}}}-1 \\
\therefore \kappa_{r_h}(n_u) &= \mathrm{e}^{\mathrm{j}\varphi_{\mathrm{R}_h}(u)}\cdot \mathrm{e}^{-\mathrm{j}2\pi iN_{\mathrm{ch}}\Delta f\tau_h}.
\end{aligned} \tag{2.26}$$

It can be seen that the range phase rotation term can be split into two terms with the first phase term due to the user index u affecting all modulation symbols of the same frame by a constant amount, and the second phase term rotating over the subcarrier set n_u along the frequency-subcarrier axis. The vector $\vec{\kappa}_{r_h}$, which is the phase rotation term over n_u subcarriers can then be written as

$$\begin{aligned}
\vec{\kappa}_{r_h} &= \begin{bmatrix}0 & \mathrm{e}^{\mathrm{j}\varphi_{\mathrm{R}_h}(u)}\mathrm{e}^{-\mathrm{j}2\pi N_{\mathrm{ch}}\Delta f\tau_h} & \dots & \mathrm{e}^{\mathrm{j}\varphi_{\mathrm{R}_h}(u)}\mathrm{e}^{-\mathrm{j}2\pi(\frac{N}{N_{\mathrm{ch}}}-1)N_{\mathrm{ch}}\Delta f\tau_h}\end{bmatrix}^{\mathrm{T}} \\
&= \mathrm{e}^{\mathrm{j}\varphi_{\mathrm{R}_h}(u)}\begin{bmatrix}0 & \mathrm{e}^{-\mathrm{j}2\pi N_{\mathrm{ch}}\Delta f\tau_h} & \dots & \mathrm{e}^{-\mathrm{j}2\pi(\frac{N}{N_{\mathrm{ch}}}-1)N_{\mathrm{ch}}\Delta f\tau_h}\end{bmatrix}^{\mathrm{T}}.
\end{aligned} \tag{2.27}$$

The effect of these two terms are plotted in Fig. 2.5 for a target located at $r = 50\,\mathrm{m}$ with $\Delta f = 90.9\,\mathrm{kHz}$. It can be seen that $\mathrm{e}^{\mathrm{j}\varphi_{\mathrm{R}_h}(u)}$ adds a fixed phase offset to all modulation symbols on the same subcarrier set governed by u, while the second term rotates over the frequency-subcarrier axis for all frames reflected from the same target. In the estimation of the range, only the second term is of importance. This is because a fixed phase addition to all subcarriers will not affect relative phase difference between the subcarriers, a criterion for the radar's range estimation.

Effect of target's Doppler

Here, it is of interest to see how the Doppler phase term in time domain affects the modulation symbol in frequency domain. It has been shown in the derivation of Eq. (2.11) that the Doppler term rotates through time t. Thus the translation of t to the OFDM frame at the baseband level (after the serial-to-parallel conversion) in terms of the duration per OFDM symbol T must first be investigated. As has been explained in Section 2.1.1, every OFDM symbol is also prepended with a CP, making it longer to result in $T = T_0 + T_{\text{CP}}$. Despite this, the sampling time T_S remains and hence can be written as

$$T_\text{S} = \frac{T_0}{N} = \frac{1}{N\Delta f} \quad \text{or} \quad T_\text{S} = \frac{T}{N + N_\text{CP}}. \tag{2.28}$$

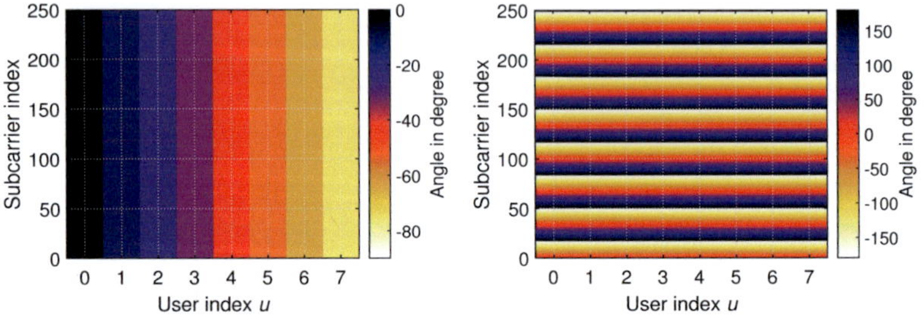

(a) Range phase rotation imposed by the user index.

(b) Range phase rotation over subcarriers.

Fig. 2.5: Phase rotation due to range and user index terms at $r = 50\,\text{m}$ and $\Delta f = 90.9\,\text{kHz}$.

Consider also a general case, where training symbols (also called preambles) are prepended to the OFDM transmit frame for synchronization and payload data recovery (discussed in Chapter 4). The total length of the training symbols in time domain is denoted as $N_{\text{train,tot}} = M_\text{train} \cdot (N_\text{train} + N_{\text{train,CP}})$, where M_train and $(N_\text{train} + N_{\text{train,CP}})$ are analogous to M and $(N + N_\text{CP})$ of the radar symbol matrix respectively. After the serial-to-parallel and IDFT operations at the Rx, the time term t must be rewritten to match the time incurred by the OFDM frame size of $\mathbb{C}^{(N+N_\text{CP})\times M}$. Using the sampling time defined in Eq. (2.28) and the Doppler term from Eq. (2.11), the t term can be translated to the modulation symbol domain in the following way

$$\begin{aligned}
\kappa_{\text{D}_h}(n,m) &= \mathrm{e}^{\mathrm{j}2\pi f_{\text{D},h} T_\text{S}\left(n + m(N+N_\text{CP}) + N_\text{train,tot}\right)} \\
&= \mathrm{e}^{\mathrm{j}2\pi f_{\text{D},h} n T_\text{S}} \cdot \mathrm{e}^{\mathrm{j}2\pi f_{\text{D},h} m (N+N_\text{CP}) T_\text{S}} \cdot \mathrm{e}^{\mathrm{j}2\pi f_{\text{D},h} N_\text{train,tot} T_\text{S}} \\
&= \mathrm{e}^{\mathrm{j}2\pi f_{\text{D},h} n T_\text{S}} \cdot \mathrm{e}^{\mathrm{j}2\pi f_{\text{D},h} m T} \cdot \mathrm{e}^{\mathrm{j}2\pi f_{\text{D},h} M_\text{train} T} \\
&= \mathrm{e}^{\mathrm{j}\varphi_{\text{D}_h}(n)} \cdot \mathrm{e}^{\mathrm{j}2\pi f_{\text{D},h} m T} \cdot \mathrm{e}^{\mathrm{j}\varphi_\text{train}}.
\end{aligned} \tag{2.29}$$

Fig. 2.6 shows the effect of Doppler rotation over the frequency-subcarrier and time-symbol axes for a small ($f_D T_0 = 0.00495$) and large ($f_D T_0 = 0.44$) Doppler shift respectively. The frequency offset is usually expressed as a ratio of the offset to the subcarrier spacing hence $f_D/\Delta f = f_D T_0$. It can be seen that the term $\mathrm{e}^{\mathrm{j}\varphi_{D_h}(n)}$ rotates over every subcarrier in Fig. 2.6(a) and 2.6(b). They are however very small when compared to the phase rotation caused by the term $\mathrm{e}^{\mathrm{j}2\pi f_{D,h} mT}$ shown in Fig. 2.6(c) and 2.6(d). The overall effect of the $\kappa_{D_h}(n,m)$ term is depicted in Fig. 2.6(e) and 2.6(f).

The term $\mathrm{e}^{\mathrm{j}\varphi_{\mathrm{train}}}$ adds a constant phase shift over all modulation symbols just like $\mathrm{e}^{\mathrm{j}\varphi_{R_h}(u)}$, hence this term does not affect the Doppler estimation. It must be pointed out that the $\mathrm{e}^{\mathrm{j}\varphi_{D_h}(n)}$ term actually causes Inter-Carrier Interference (ICI), the consequence of the OFDM subcarriers losing their orthogonality. For this reason, the parameterization for radar use is proposed in [Stu12] to allow for the OFDM subcarrier spacing to be at least ten times larger than the expected maximum Doppler frequency in the scenario, $\Delta f \gg 10 f_{D,\max}$. When used in a multi-user access scenario, this condition is however also subjected to the signal-to-interference ratio seen at the input of the Rx ($\mathrm{SIR_{in}}$) and hardware frequency offset and will be explained in Chapter 4.

Supposedly there are H number of targets, each moving with an arbitrary velocity. The Rx signal will then be composed of an overlay of these Doppler shifted signals to result in a complex phase rotation term over the frequency-subcarrier and time-symbol axes. The Doppler terms can add up constructively and destructively depending on the direction. Taking into consideration that κ_{r_h} has a more pronounced effect than $\mathrm{e}^{\mathrm{j}\varphi_{D_h}(n)}$ on the frequency-subcarrier axis, it would hence be more rational to utilize the $\mathrm{e}^{\mathrm{j}2\pi f_{D,h} mT}$ term to estimate the Doppler shift. It is also of interest to see how the $\mathrm{e}^{\mathrm{j}\varphi_{D_h}(n)}$ term interacts with the κ_{r_h} term. Rewriting the term $\mathrm{e}^{\mathrm{j}\varphi_{D_h}(n)}$ to match n_u,

$$\mathrm{e}^{\mathrm{j}\varphi_{D_h}(n_u)} = \mathrm{e}^{\mathrm{j}2\pi f_{D,h} n_u T_S} = \mathrm{e}^{\mathrm{j}2\pi(u+iN_{\mathrm{ch}}) f_{D,h} \frac{T_0}{N}}. \tag{2.30}$$

To see its effect on the range term, it is multiple with κ_{r_h} and by manipulating $\Delta f = \frac{1}{T_0}$ and T_S to obtain

$$\mathrm{e}^{\mathrm{j}\varphi_{D_h}(n_u)} \cdot \kappa_{r_h}(n_u) = \mathrm{e}^{-\mathrm{j}2\pi(u+iN_{\mathrm{ch}})\left(\frac{N\tau_h - f_{D,h} T_0^2}{N T_0}\right)}. \tag{2.31}$$

For radar applications in the giga-hertz spectrum, the T_0 is in the range of microseconds while the Doppler shift is in the kilohertz range. Hence the term

$f_{D,h}T_0^2 \approx 0$ rendering $e^{j\varphi_{D_h}(n_u)} \approx 1$, which means this term has minimal effect on the range estimation. Taking this into account and rewriting the $\kappa_{D_h}(n_u, m)$ term as a vector yields

$$\vec{\kappa}_{D_h} = e^{j\varphi_{\text{train}}} \begin{bmatrix} 0 & e^{j2\pi f_{D,h}T} & \dots & e^{j2\pi f_{D,h}(M-1)T} \end{bmatrix}. \tag{2.32}$$

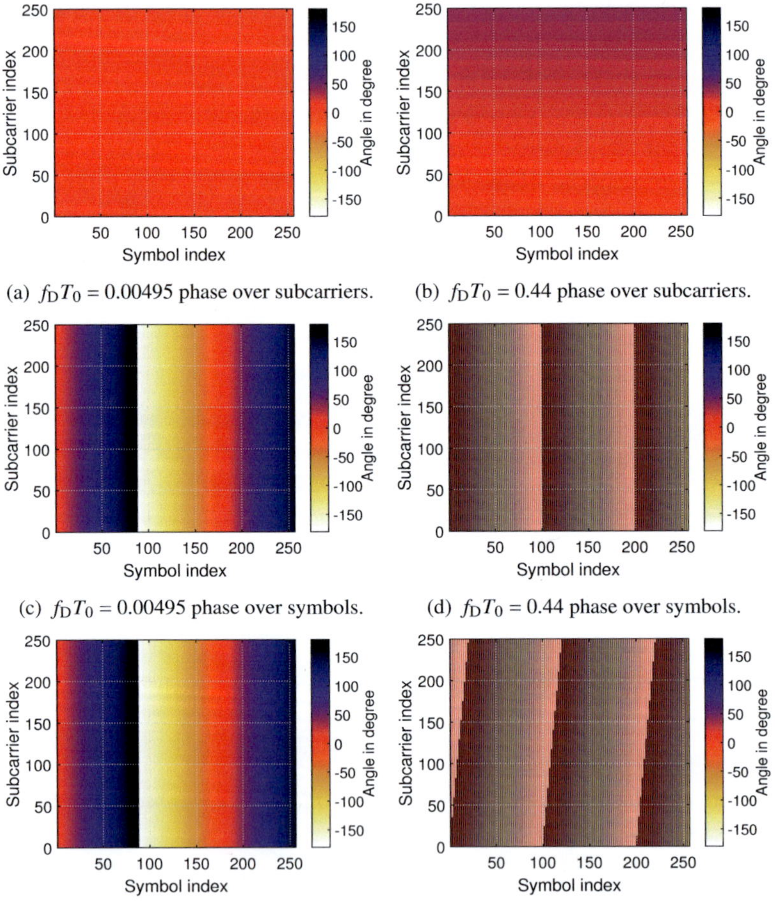

(a) $f_D T_0 = 0.00495$ phase over subcarriers.

(b) $f_D T_0 = 0.44$ phase over subcarriers.

(c) $f_D T_0 = 0.00495$ phase over symbols.

(d) $f_D T_0 = 0.44$ phase over symbols.

(e) $f_D T_0 = 0.00495$ total Doppler phase.

(f) $f_D T_0 = 0.44$ total Doppler phase.

Fig. 2.6: Phase rotation due to Doppler term for $f_D T = 0.00495$ and $f_D T = 0.44$ over, - 1) frequency-subcarrier axis, 2) time-symbol axis 3) total frame.

2.4.3 Receive signal

At the Rx side, the baseband modulation symbol is stripped of the training symbols and CP. Taking into account all channel state information (CSI) terms analogous to Eq. (2.11) and the range and Doppler terms in Eq. (2.27) and Eq. (2.32) respectively, the Rx signal matrix can be expressed as

$$\mathbf{Y}_q(n,m) = \sum_{p=0}^{P-1} \mathbf{X}_p(n_u,m) \sum_{h=0}^{H-1} \alpha_{p,q,h}\, \mathrm{e}^{\mathrm{j}(\Psi_{p,q,h}+\zeta_{n_u,m})} \left(\mathrm{e}^{\mathrm{j}\varphi_{\mathrm{R}_h}(u)}\, \mathrm{e}^{-\mathrm{j}2\pi(iN_{\mathrm{ch}})\Delta f \tau_{p,q,h}}\right) \cdot \left(\mathrm{e}^{\mathrm{j}2\pi f_{\mathrm{D},p,q,h} mT}\, \mathrm{e}^{\mathrm{j}\varphi_{\mathrm{train}}}\right) + \hat{\mathbf{Z}}(n,m), \tag{2.33}$$

where it can be seen that in the ideal case, each Rx receives non-overlapping signal contributions from all Tx (as long as all p are assigned independent u). The receive matrix at every Rx has a full matrix of size of $\mathbb{C}^{N\times M}$. Now, a description of the terms is in order to attempt to simplify the equation of the Rx signal as well as to give an intuitive understanding to these terms, which will be frequently invoked from now on.

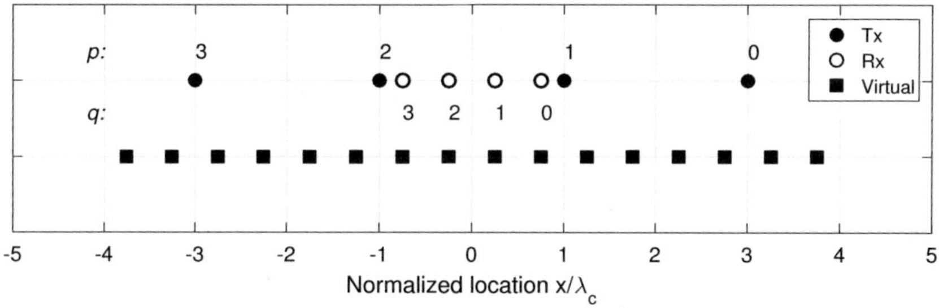

Fig. 2.7: The simplest physical placements of a linear 1D 4×4 antenna array with the resulting 16-element virtual array. It will be discussed in more detail in Chapter 3.

As will be explained in Section 3.2.1, the geometry of the physical transmit and receive array antennas in the simplest case is the one-dimensional (1D) array with $P = Q = 4$ is as depicted in Fig. 2.7. All antenna elements lie on the x-axis. The subscripts of p and q for some terms in Eq. (2.33) can be simplified while some cannot, based on a few justifications as follows:

- The round-trip time for all qp-th paths is given by

$$\begin{aligned} \tau_{p,q,h} &= \frac{r_{p,h}+r_{q,h}}{c_0}, \qquad r_{p,h} \approx r_{q,h} \approx r_h \\ \Rightarrow \tau_{p,q,h} \approx \tau_h &= \frac{2r_h}{c_0}, \qquad \text{for all } p,q, \end{aligned} \tag{2.34}$$

  where $r_{p,h}$ and $r_{q,h}$ are the distances from the p-th or q-th antenna to the target h respectively. These distances are minuscule and can each be approximated to be r_h, which is the distance from the origin of the antenna geometry to the target h. This leads to the simplification where the range estimation is the same for all transmit-receive pairs based on the round-trip time τ_h .

- The relative velocity estimated by all receivers is actually the absolute value of the parallel velocity vector $\vec{v}_{||,p,q,h}$. Hence in principle, all transmit-receive antenna pairs will 'see' a slightly different amount of Doppler phase rotation, where

$$f_{\mathrm{D},p,q,h} = \frac{2v_{\mathrm{rel},p,q,h}}{\lambda_\mathrm{c}}, \qquad v_{\mathrm{rel},p,q,h} = |\vec{v}_{||,p,q,h}|. \tag{2.35}$$

  Thus, the $f_{\mathrm{D},p,q,h}$ can be reduced to $f_{\mathrm{D},h}$ for all transmit-receive pairs only if the spacing between the array elements is much smaller than the euclidean distance of the moving target from the center of the 1D array, r_h. Considering the transmit and receive array element spacing to be d_T and d_R respectively, the largest distance between the transmit antenna on one end to the furthest receive antenna on the other end of the array according to Fig. 2.7 is from $p = 0$ to $q = 3$ or vice versa. Let x_{T_p} and x_{R_q} be the x-coordinate location of the elements for the p-th Tx antenna and q-th Rx antenna respectively. The separation between the first Tx antenna and the last Rx antenna is given by

$$\begin{aligned} x_{\mathrm{T}_0} - x_{\mathrm{R}_Q} &= \frac{1}{2}\left[x_{\mathrm{T}_0} - x_{\mathrm{T}_P}\right] + \frac{1}{2}\left[x_{\mathrm{R}_0} - x_{\mathrm{R}_Q}\right] \\ &= \frac{d_\mathrm{T}}{2}(P-1) + \frac{d_\mathrm{R}}{2}(Q-1), \end{aligned} \tag{2.36}$$

  with the Tx antenna element spacing $d_\mathrm{T} = Q \cdot d_\mathrm{R}$ and $d_\mathrm{R} = \frac{\lambda_\mathrm{c}}{2}$ to avoid grating lobes to result in $x_{\mathrm{T}_0} - x_{\mathrm{R}_3} = 3.75\lambda_\mathrm{c}$. At the operating frequency of 24 GHz this is 5 cm, which is very much less than the expected target distance from $+\Delta r$ onward. Since this radar is not designed to be an ultra-precision near-range radar application, it can be safely assumed that the nearest object distance to the radar that can be properly measured must be more than one range resolution bin. Hence the term $f_{\mathrm{D},p,q,h}$ can be reduced to $f_{\mathrm{D},h}$ for all transmit-receive pairs.

- for the same reason as above, the amplitude attenuation can also be reduced from $\alpha_{p,q,h}$ to α_h.
- The term $\mathrm{e}^{\mathrm{j}\Psi_{p,q,h}}$ can be expanded as

$$\Psi_{p,q,h} = -2\pi \frac{|r_{p,h} + r_{q,h}|}{\lambda_{\mathrm{c}}}, \qquad \text{where } |r_{p,h} + r_{q,h}| \approx |2r_h| \gg \lambda_{\mathrm{c}}. \tag{2.37}$$

  This term cannot be simplified because each transmit-receive pair will acquire a slightly different phase term, which is constant over the whole qp-th frame. This term is essential in DOA estimation and will be discussed in Chapter 3.

- The term $\mathrm{e}^{\mathrm{j}\zeta_{n_u,m}}$ depends on the subcarrier and symbol indices since this term is mostly hardware dependent. When there is a less than perfect calibration at the multiple Tx and Rx ports of the hardware, this term will appear with a certain phase offset or rotation over the frequency-subcarrier and time-symbol axes. This will then result in imperfect range and Doppler estimation. If the hardware has been calibrated, this term can be neglected.

Taking into consideration all the aforementioned points and justifications, and assuming that the hardware instrumentation of the RadCom has been calibrated (thus $\mathrm{e}^{\mathrm{j}\zeta_{n_u,m}} = 1$), the received symbols at the q-th receiver can be written as

$$\begin{aligned} \mathbf{Y}_q(n,m) = \sum_{p=0}^{P-1} \mathbf{X}_p(n_u,m) \sum_{h=0}^{H-1} \alpha_h \mathrm{e}^{\mathrm{j}\Psi_{p,q,h}} \Big(\mathrm{e}^{\mathrm{j}\varphi_{\mathrm{R}_h}(u)} \, \mathrm{e}^{-\mathrm{j}2\pi(iN_{\mathrm{ch}})\Delta f \tau_h} \Big) \cdot \\ \Big(\mathrm{e}^{\mathrm{j}\varphi_{\mathrm{train}}} \, \mathrm{e}^{\mathrm{j}2\pi f_{\mathrm{D},h} m T} \Big) + \hat{\mathbf{Z}}(n,m). \end{aligned} \tag{2.38}$$

The size of the $\mathbf{Y}_{qp}$ matrix (q-th Rx matrix due to the p-th Tx) without the unused subcarriers will shrink to the size $\mathbb{C}^{\frac{N}{N_{\mathrm{ch}}}\times M}$. The $\mathbf{Y}_{qp}$ matrix is expressed as

$$\begin{aligned}\mathbf{Y}_{qp}(n_u,m) &= \mathbf{X}_p(n_u,m)\sum_{h=0}^{H-1}\alpha_h \mathrm{e}^{\mathrm{j}\Psi_{p,q,h}}\left(\mathrm{e}^{\mathrm{j}\varphi_{\mathrm{R}_h}(u)}\,\mathrm{e}^{-\mathrm{j}2\pi(iN_{\mathrm{ch}})\Delta f\tau_h}\right)\cdot \\ &\qquad\left(\mathrm{e}^{\mathrm{j}\varphi_{\mathrm{train}}}\,\mathrm{e}^{\mathrm{j}2\pi f_{\mathrm{D},h}mT}\right)+\hat{\mathbf{Z}}(n_u,m) \\ &= \mathbf{X}_p(n_u,m)\sum_{h=0}^{H-1}\alpha_h \overbrace{\mathrm{e}^{\mathrm{j}\Psi_{p,q,h}}}^{\text{DOA}}\cdot\ \overbrace{\vec{\kappa}_{r_h}\otimes\vec{\kappa}_{\mathrm{D}_h}}^{\text{range \& Doppler}}\ +\hat{\mathbf{Z}}(n_u,m),\end{aligned} \tag{2.39}$$

where [⊗] signifies the operator for the outer product.

2.5 Range and Doppler estimation

The OFDM signal model is basically composed of sinusoids hence the estimation problem becomes, according to [Bra14], one that estimates phases or frequencies of the sampled complex sinusoids. These sinusoids are spectrally superpositioned based on their indices at the Rx. This means that instead of the conventional cross-correlation method (as shown in Eq. (2.40)) in time domain, the range and Doppler estimation can be performed on the frequency domain signal. While parametric methods such as ESPRIT [RPK86] and MUSIC [Sch86] can be applied, a periodogram-based spectral estimation method proposed in [Stu12] and elaborated in [Bra14] is used throughout this thesis.

$$f_{r_h,v_{\mathrm{rel},h}}(\tau_h,f_{\mathrm{D},h}) = \int_{-\infty}^{\infty} x^*(t-\tau_h)\cdot y(t)\,\mathrm{e}^{-\mathrm{j}2\pi f_{\mathrm{D},h}t}\,dt \tag{2.40}$$

It has been seen in [Stu12] that the use of cross-correlation (XCORR) causes spurious sidelobes due to the XCORR's dependency on the amplitude of the modulated data and the strength of the correlation term between the transmit and receive signal. When the sidelobes are too high they can mask the existence of an estimated object leading to ambiguity - a false positive or negative result. The periodogram method on the other hand uses only Fourier transforms, where the resulting sidelobes will only be due to the transform itself. This method is elaborated in the following section.

2.5.1 Fourier Transform-based estimation

To estimate the range and relative velocity of the objects, only the terms τ_h and $f_{D,h}$ are needed. The term α_h only affects the amplitude of the Rx signal (in terms of signal-to-noise ratio, SNR) while $e^{j\Psi_{p,q,h}}$ is constant over all modulation symbols of the qp-th frame hence has no effect on the estimation outcome. The noise term $\mathbf{Z}(n,m)$ also influences the SNR of the receive signal. Assuming that the SNR is sufficiently high, the known transmit modulation symbols can be removed from the receive signal by means of an element-wise division. Since each Rx receives signals from all Tx, the quotient matrix $\mathbf{D}_{qp}(n_u,m)$ depends on the the transmit symbols of the p-th Tx transmitting on n_u, $\mathbf{X}_p(n_u,m)$. Thus there will be $P \cdot Q$ number of quotient matrices available for evaluation at the radar processing block. The quotient elements based on the element-wise division of receive by transmit symbols is

$$\begin{aligned}(D_{qp})_{n_u,m} &= \frac{(Y_{qp})_{n_u,m}}{(X_p)_{n_u,m}} \\ &= \sum_{h=0}^{H-1} \alpha_h \mathrm{e}^{\mathrm{j}\Psi_{p,q,h}} \left(\mathrm{e}^{\mathrm{j}\varphi_{\mathrm{R}_h}(u)}\, \mathrm{e}^{-\mathrm{j}2\pi(iN_{\mathrm{ch}})\Delta f \tau_h}\right)\left(\mathrm{e}^{\mathrm{j}\varphi_{\mathrm{train}}}\, \mathrm{e}^{\mathrm{j}2\pi f_{\mathrm{D},h} mT}\right) + (Z_{qp})_{n_u,m},\end{aligned} \tag{2.41}$$

where the Gaussian white noise term $\frac{(\hat{Z}_{qp})_{n_u,m}}{(X_p)_{n_u,m}} = (Z_{qp})_{n_u,m}$ is still white. Thus the elements of the quotient matrix in Eq. (2.41) can be rewritten as

$$\mathbf{D}_{qp}(n_u,m) = \sum_{h=0}^{H-1} \alpha_h \mathrm{e}^{\mathrm{j}\Psi_{p,q,h}} \left(\vec{\kappa}_{r_h} \otimes \vec{\kappa}_{\mathrm{D}_h}\right) + \mathbf{Z}(n_u,m). \tag{2.42}$$

Since the $\vec{\kappa}_{r_h}$ and $\vec{\kappa}_{\mathrm{D}_h}$ terms affect the different axes and are thus orthogonal to each other, the time delay and the Doppler terms can be estimated directly with Eq. (2.42). Demonstrated in the following are the range and Doppler estimation procedures done separately on the quotient matrix $\mathbf{D}_{qp}(n_u,m)$.

Range estimation

The range profile is estimated by taking an IDFT through the frequency-subcarrier axis given by

$$\begin{aligned}\mathbf{R}(k,m) &= \mathrm{IDFT}\left[\mathbf{D}_{qp}(n_u,m)\right]\\ &= \frac{1}{N}\sum_{i=0}^{\frac{N}{N_{\mathrm{ch}}}-1}\mathbf{D}_{qp}(n_u,m)\cdot \mathrm{e}^{\mathrm{j}2\pi\frac{i}{N/N_{\mathrm{ch}}}k}\\ &= \frac{1}{N}\sum_{h=0}^{H-1}\alpha_h \mathrm{e}^{\mathrm{j}\Psi_{p,q,h}}\,\mathrm{e}^{\mathrm{j}\varphi_{\mathrm{R}_h}(u)}\cdot\sum_{i=0}^{\frac{N}{N_{\mathrm{ch}}}-1}\left[\mathrm{e}^{-\mathrm{j}2\pi i N_{\mathrm{ch}}\Delta f\tau_h}\,\mathrm{e}^{\mathrm{j}2\pi\frac{i}{N/N_{\mathrm{ch}}}k}\right]\cdot\vec{\kappa}_{\mathrm{D}_h}.\end{aligned} \tag{2.43}$$

Equating both exponential terms in the brackets, the maxima for the h-th object will occur at the indices $\hat{k}_h$ where

$$\begin{aligned}\hat{k}_h &= N\Delta f\tau_h\\ &= \frac{2N\Delta f}{c_0}r_h, \qquad \hat{k}_h = 0,1,\ldots,\frac{N}{N_{\mathrm{ch}}}-1.\end{aligned} \tag{2.44}$$

Each h-th object will give rise to one peak at $\hat{k}_h \in \mathbb{Z}$ along the frequency-subcarrier axis. Obtaining the equivalent range r_h is a matter of multiplying the index $\hat{k}_h$ with $c_0/(2N\Delta f)$. The amplitude of the peak is dependent on the energy that is backscattered by the h-th object and is related to its RCS and angle of orientation toward the radar.

Relative velocity estimation

The relative velocity profile is estimated by taking a DFT over the symbols as given by

$$\begin{aligned}\mathbf{V}(n_u,l) &= \mathrm{DFT}\left[\mathbf{D}_{qp}(n_u,m)\right]\\ &= \sum_{m=0}^{M-1}\mathbf{D}_{qp}(n_u,m)\cdot \mathrm{e}^{-\mathrm{j}2\pi\frac{m}{M}l}\\ &= \sum_{h=0}^{H-1}\alpha_h \mathrm{e}^{\mathrm{j}\Psi_{p,q,h}}\,\mathrm{e}^{\mathrm{j}\varphi_{\mathrm{train}}}\cdot\sum_{m=0}^{M-1}\left[\mathrm{e}^{\mathrm{j}2\pi f_{\mathrm{D},h}mT}\,\mathrm{e}^{-\mathrm{j}2\pi\frac{m}{M}l}\right]\cdot\vec{\kappa}_{r_h}.\end{aligned} \tag{2.45}$$

Just like with the range profile, each h-th object will give rise to one peak at $\hat{l}_h \in \mathbb{Z}$ along the time-symbol axis as given in Eq. (2.46). Multiplying the index $\hat{l}_h \cdot c_0/(2MTf_c)$ will yield the $v_{\mathrm{rel},h}$.

$$\begin{aligned} \hat{l}_h &= M f_{\mathrm{D}} T \\ &= \frac{2MTf_{\mathrm{c}}}{c_0} v_{\mathrm{rel},h}, \qquad \hat{l}_h = 0, 1, ..., M-1 \end{aligned} \tag{2.46}$$

Processed radar matrix

The processed radar matrix used to plot the radar image[8] is then given by,

$$\mathbf{I}_{qp}(k,l) = \mathrm{IDFT}_{n_u}\Big[\mathrm{DFT}_m\Big[\mathbf{D}_{qp}(n_u,m)\Big]\Big]. \tag{2.47}$$

The range and Doppler profiles are estimated in the sequence of the IDFT then DFT or vice versa. The order of the estimation sequence will not affect the outcome of the radar estimation.

Processing gain

Due to the DFT and IDFT operations, the SNR of processed radar matrix $\mathbf{I}_{qp}$ is increased by a factor of $\frac{N}{N_{\mathrm{ch}}} \cdot M$ for each qp-th received frame. This is also called the 'processing gain' and is denoted as G_{per} (gain due to the periodogram method) as given in Eq. (2.48). Due to the spectrally interleaved signal model, like the $r_{\mathrm{max,interleaved}}$ the G_{per} is also affected and is reduced by a factor of N_{ch} compared to the case of a SISO radar where all N subcarriers are used by the single Tx antenna.

$$G_{\mathrm{per}} = \frac{N}{N_{\mathrm{ch}}} \cdot M \tag{2.48}$$

Despite the lower processing gain, the total transmit power of the spectrally interleaved signal (assuming the transmit power is the same as the SISO case) will be divided among fewer subcarriers. This means that the power on every subcarrier also increases by a factor of N_{ch} compared to the classical OFDM

[8] featuring range and Doppler

signal model, where all N subcarriers are used. Hence the overall SNR of the signal bandwidth will remain the same.

2.5.2 Radar output quality metric

A metric for judging the quality of the radar estimation from the radar image is the '**Signal-to-Noise ratio** of the processed radar matrix ($\mathrm{SNR_{out}}$)'. In the presence of interferers the metric '**Signal-to-Noise plus Interferer ratio** at the output of the radar ($\mathrm{SNIR_{out}}$)' is used instead. This is simply the ratio of the highest peak in the radar image due to the nearest object to the average noise floor (and interferer). The $\mathrm{SNR_{out}}$ and $\mathrm{SNIR_{out}}$ encompass the SNR at the receiver, the processing gain, any windowing loss and the average noise floor due to the background noise (not inclusive of the sidelobes caused by the periodogram estimation method). $\mathrm{SNR_{out}}$ and $\mathrm{SNIR_{out}}$ are hence dimensionless and are often given in terms of dB.

2.5.3 Zero-padding and DFT length

The quotient matrix $\mathbf{D}_{qp}(n_u, m)$ is often zero-padded before the IDFT and DFT procedures to result in a finer estimation step in the range and Doppler processed radar matrix. Plotting the zero-padded matrix also provides for better visualization of the detected objects. With zero-padding, the matrix size of $\mathbf{I}_{qp}(k,l)$ will grow to $\mathbb{C}^{k \cdot z_{\mathrm{pad},k} \times l \cdot z_{\mathrm{pad},l}}$, where $z_{\mathrm{pad}} \in \mathbb{Z}$ is the zero-padding factor. It must be stressed here that zero-padding will not increase the resolution. The one bin that represent the range or velocity resolution is only further divided into z_{pad} bins. This then helps improve scalloping loss or straddle loss that is caused by the regular grid of the Fourier transform-based processing elaborated in [Bra14, pp. 47–51].

2.5.4 Shadowing effect

The shadowing effect is a classic radar problem, which is best illustrated with two examples. First, when a target with a large RCS is located near the radar, other targets with relatively smaller RCS will not show up on the radar image due to its limited $\mathrm{SNR_{out}}$ or $\mathrm{SNIR_{out}}$. Second, when a target is hidden by an obstacle, the target also will not appear on the radar image. This effect is called ‘shadowing’. To avoid this some radars are tilted to look down at an angle so that no objects are hidden from view. By the same reasoning, any antenna coupling must also be calibrated or removed else real targets might be overshadowed leading to a false negative radar estimation.

2.5.5 Windowing losses

The processed radar matrix can be further improved using windowing to suppress the sidelobes even further but at the expense of the radar image’s resolution and $\mathrm{SNR}_{\mathrm{out}}$ of the radar image. It is common knowledge that a window function with better sidelobe attenuation has a larger mainlobe. Since each type of window function comes with varying mainlobe width, height of peaks and sidelobe attenuation, it will cause different amount of loss of $\mathrm{SNR}_{\mathrm{out}}$ in the periodogram. This loss is attributed to the equivalent noise bandwidth of the window, where a figure of merit for various windows is presented by Harris in [Har78]. Braun in [Bra14, ch. 3, p. 45] provides the value of the loss for various length of the Hamming window, where a window length of $N = 128$ to $N = 1024$ is shown to possess almost the same loss factor of 1.36 dB.

Throughout this thesis the Hamming window is used, once column-wise in the range estimation and again row-wise for the Doppler estimation. For a SISO OFDM model where the frame size is $N \times M$, the loss due to the Hamming windowing is approximately $1.35 + 1.36 = 2.71$ dB. This has been proven by Sturm in [Stu12, ch. 3, p. 66]. For a shorter window length due to the increment of N_{ch}, the Hamming windowing loss is taken to be approximately 2.7 dB and will be used throughout this work.

Shown in Fig. 2.8(a) are the commonly used window functions in the frequency domain representation, where the width of the main lobe versus the height of the sidelobes can be observed. Meanwhile Fig. 2.8(b) to Fig. 2.8(e) shows the radar images from one scenario processed with rectangular (boxcar), Hamming and Chebyshev (with 60 dB and 80 dB sidelobe suppression respectively) windows. The trade off between the sidelobe suppression and image resolution can be clearly seen when comparing the window functions to the rectangular window. Since windowing is not vital to the work done in this thesis, it will not be further elaborated.

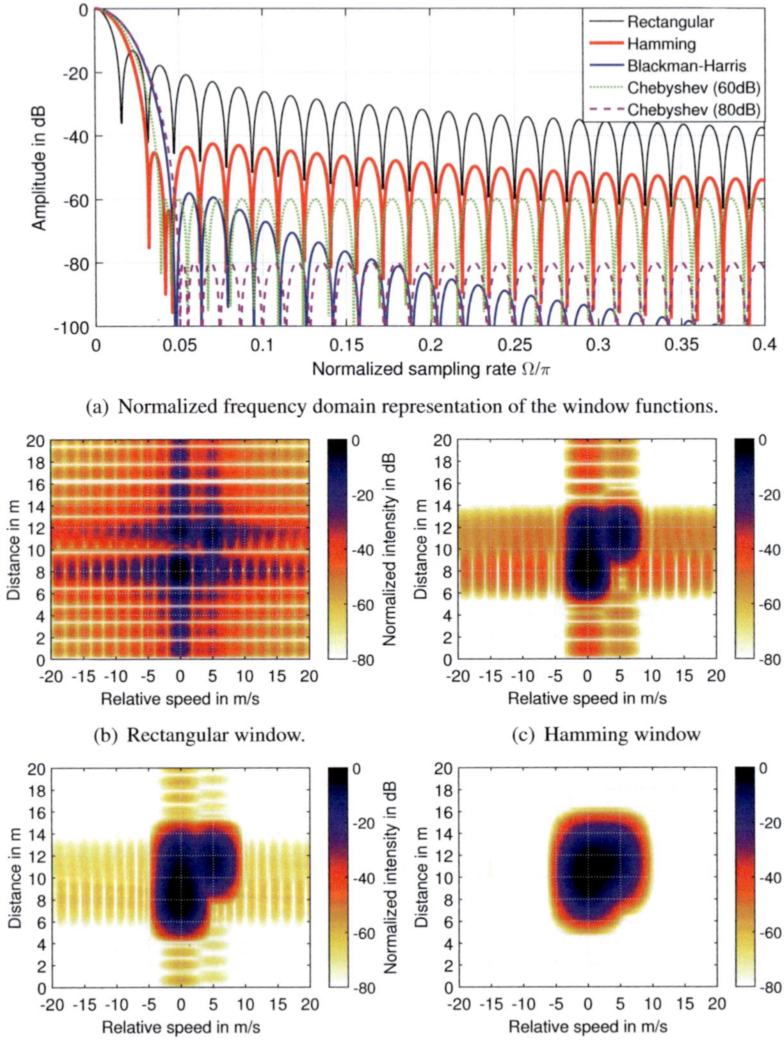

(a) Normalized frequency domain representation of the window functions.

(b) Rectangular window.

(c) Hamming window

(d) Chebyschev window, 60 dB sidelobe suppression.

(e) Chebyschev window, 80 dB sidelobe suppression.

Fig. 2.8: Effect of different window functions on the radar image with the same scenario of three targets. The target distances and relative velocities are (8 m,0 m/s), (12 m,0 m/s) and (12 m,5 m/s). All of them possess the same RCS of 5 m^2.

2.5.6 Range ambiguity due to guard interval

To complete the analysis of the signal model, it will be of interest to compare the consequence of evaluating the radar signal in the way OFDM communication signals are evaluated. In communication systems, the start point of the signal taken anywhere within the CP will not pose a problem to the correct demodulation of the payload data. For radar systems however, the effect on the range estimation can be clearly seen in Fig. 2.9, showing an autocorrelation function of an OFDM signal of $N = 400$ and $N_{\mathrm{CP}} = 50$. If the CP is not removed before the radar processing, a range ambiguity will arise at the positive time lag index equivalent to the length of the CP.

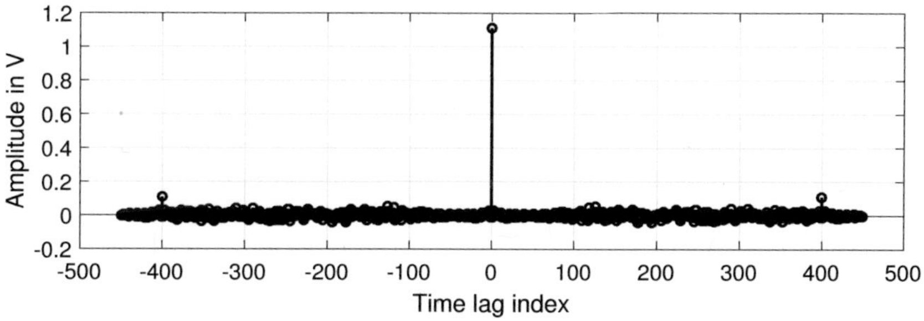

Fig. 2.9: Double-sided autocorrelation of a signal with $N = 400$ and guard interval $N_{\mathrm{CP}} = \frac{N}{8}$. The two smaller peaks on both sides of the main peak are due to the guard interval.

In practice, the erroneous removal of the CP length can happen due to hardware synchronization error between the multiple Tx and Rx. This will cause the wrong length of the CP to be removed and consequently affect the range estimation accuracy. It is thus of interest here to analyze the dependency and extent of this error on the length of the CP to be removed. This range error can be demonstrated with the following equations while referring to the notations in Fig. 2.2. To simplify the problem, assume here the case of a SISO radar where all subcarriers are used. The m-th transmit modulation symbols $a_m(n)$ is passed through the IDFT block to yield the discrete time domain samples $b_m(k)$ given by

$$b_m(k) = \sum_{n=0}^{N-1} a_m(n) \mathrm{e}^{\mathrm{j}\frac{2\pi}{N}nk}. \tag{2.49}$$

Since the CP is a copy of υ samples from the tail of the signal prepended to its start, $b_m(k)$ can be expanded as

$$\vec{b}_m = [b_m(-\upsilon) \quad b_m(-\upsilon+1) \quad \cdots \quad b_m(-1) \quad b_m(0) \quad \cdots \quad b_m(N-1)]. \quad (2.50)$$

With the added CP, $b_m(k')$ becomes

$$b_m(k') = \sum_{n=0}^{N-1} a_m(n) \mathrm{e}^{\mathrm{j}\frac{2\pi}{N} n k'}, \qquad k' = -\upsilon, ..., 0, ..., N-1. \quad (2.51)$$

The CP merely lengthens the time domain signal without any change to the sampling time. At the channel, the time domain analog signal is backscattered by one target. Taking into account all channel effects, the discrete time receive signal with the CP removed can be expressed as

$$\tilde{b}_m(\ell) = \mathrm{e}^{\mathrm{j}(\Psi+\zeta)} \, \mathrm{e}^{\mathrm{j}2\pi f_\mathrm{D} T_\mathrm{S}[\ell+mN]} \overbrace{\sum_{n=0}^{N-1} a_m(n) \mathrm{e}^{\mathrm{j}\frac{2\pi}{N} n k'}}^{\text{original transmit symbols}} \cdot \mathrm{e}^{-\mathrm{j}2\pi n \Delta f \tau} \quad (2.52)$$

$$\text{for } \ell = 0, ..., N-1.$$

The index k' in Eq. (2.52) must now take contiguous N values from the range of $\{-\upsilon, ..., 0, ..., N-1\}$ to match the Rx discrete time index ℓ. The time delay term $\mathrm{e}^{-\mathrm{j}2\pi n \Delta f \tau}$ is a frequency domain representation of the time delay in the time domain signal (c.f Eq. (2.8)) and can be written as $\mathrm{e}^{-\mathrm{j}2\pi n \frac{\tau}{T_0}}$. This term adds a phase offset to each subcarrier according to their subcarrier index without adding any additional frequency offset. The Doppler term $\mathrm{e}^{\mathrm{j}2\pi f_\mathrm{D} t}$ is a term added in the time domain and is hence dependent on the discrete time domain index of ℓ at the Rx. It is rewritten as shown in Eq. (2.52) following the convention shown in Eq. (2.29).

Taking a DFT to transform the discrete time domain signal back to frequency domain in terms of recovered modulation symbols $\tilde{a}_m(n')$, the frequency bin index is now denoted as n'. Regrouping the terms leads to

$$\begin{aligned}\tilde{a}_m(n') &= \mathrm{DFT}\{\tilde{b}_m(\ell)\}\\ &= \mathrm{e}^{\mathrm{j}(\Psi+\zeta)}\,\mathrm{e}^{\mathrm{j}2\pi f_{\mathrm{D}} mT} \underbrace{\sum_{n=0}^{N-1} \mathrm{e}^{\mathrm{j}\frac{2\pi}{N}nk'}}_{\text{IDFT term}} a_m(n)\,\mathrm{e}^{-\mathrm{j}2\pi n\Delta f\tau} \underbrace{\sum_{\ell=0}^{N-1} \mathrm{e}^{-\mathrm{j}\frac{2\pi}{N}n'\ell}}_{\text{DFT term}} \cdot \mathrm{e}^{\mathrm{j}\frac{2\pi}{N}\ell f_{\mathrm{D}} T_{\mathrm{O}}}.\end{aligned} \tag{2.53}$$

To see only the effects of the CP, the Doppler term is set to $f_{\mathrm{D}} = 0$ to avoid the loss of orthogonality. In this way, the incoming samples with the index n will align with the DFT bins of the Rx so that $n' = n$. If the CP is correctly removed with $k = \ell$, the IDFT and DFT term will vanish as given by the condition

$$\sum_{n=0}^{N-1} \mathrm{e}^{\mathrm{j}\frac{2\pi}{N}nk'} \sum_{\ell=0}^{N-1} \mathrm{e}^{-\mathrm{j}\frac{2\pi}{N}n'\ell} = N, \qquad \text{if } n = n' \text{ and } k' = \ell. \tag{2.54}$$

The symbol-under-evaluation $\tilde{a}_m(n)$ can be observed by letting $n' = n$ while retaining the k and ℓ indices, and is expressed as

$$\tilde{a}_m(n') = a_m(n)\,\mathrm{e}^{\mathrm{j}(\Psi+\zeta)} \underbrace{\mathrm{e}^{-\mathrm{j}2\pi n\Delta f\tau} \sum_{n'=0}^{N-1} \sum_{\ell=0}^{N-1} \mathrm{e}^{-\mathrm{j}\frac{2\pi}{N}n'(k'-\ell)}}_{\text{range terms}}. \tag{2.55}$$

It can be seen that the recovered symbol contains the original symbol $a_m(n)$, along with all the phase rotations associated with the channel, as well as the DFT term. For a correct demodulation, k' must start from $-\upsilon$ at the very least hence the range of values given by

$$-\upsilon \le (k' - \ell) \le 0, \qquad (k' - \ell) \in \mathbb{Z}. \tag{2.56}$$

This shows that when $(k' - \ell) \neq 0$, there will be an additional phase rotation term along the frequency-subcarrier axis that will affect the range phase rotation term. With every unit of $(k - \ell)$, the estimated range will have an error of one range resolution cell, or $+\Delta r$.

2.6 Chapter 2 summary

Like the SISO OFDM RadCom, with the MIMO OFDM RadCom, arbitrary payload data can be transmitted over each Tx antenna using independent sets of subcarriers. The Rx signal will experience phase rotations over the frequency-subcarrier axis due to the range of the target, and over the time-symbol axis due to the Doppler of the targets. The main difference between the SISO and MIMO configuration is the multiple antennas that are spaced equally apart to take advantage of the spatial diversity. Each qp-th antenna combination will receive backscattered signals of a slightly different relative phase $e^{j\Psi_{p,q,h}}$, which is constant over the qp-th frame. This very phase difference between at qp-th frame enables the MIMO configuration to estimate the direction of arrival (DOA), which is relative to the radar's angular planes (azimuth or elevation). The aforementioned range, Doppler and DOA terms are quasi-independent from each other and hence can be separately estimated.

Besides the three terms above, the user index u, that is used to define the subcarrier allocation to all P transmit antennas will introduce an extra constant phase shift over the received qp-th frame. The influence of this term will be explored further in Chapter 6.

3 Antenna array geometry for digital beamforming

While the signal model influences the range and Doppler estimation characteristics, the angular estimation is directly affected by the antenna array used. Due to the characteristics of the physical antennas, all radars and even communication systems have a certain limited angular field-of-view (FOV). For communications, in most cases an omni-directional FOV is desired. When the terrain must be taken into account however, directional antennas are utilized to avoid clutter and to focus the signal strength to certain areas/users. For localization purposes in the case of radars, only directional antennas are used else the radar cannot tell the direction where the signal is backscattered from, thus causing ambiguity. In order for the sensor e.g. airport radar, to scan 360° of its surrounding, either a mechanical turning platform or beamforming techniques to focus the radiation to a certain direction is used.

MIMO radars use the concept of digital beamforming at the receive side. Multiple transmit antennas will illuminate the area of observation within their FOV simultaneously. At the digital signal processing block of the Rx the multiple transmit signals are then distinguished and resolved by each Rx. This causes the effect as though the radiation pattern of the transmit antennas are sharpened and focused at all directions simultaneously. This phenomena can be explained by the geometry of the physical antenna arrays and the virtual array concept.

Since there are not much literature touching on the concept of virtual antennas available, this chapter serves to provide the fundamentals, which can then be expanded to more complicated and optimized antenna array structures for MIMO radars. The first part of this chapter will introduce the virtual array concept qualitatively and quantitatively. The second part of this chapter then deals with the derivation of the beamsteering vectors based on the physical geometry of the antenna arrays, which are used for direction-of-arrival (DOA) estimation of the targets.

The real antennas are called the '**physical antennas/ array/ aperture**' while the '**virtual antennas/ array/ aperture**' are the manifestation of the physical antennas due to their arrangement or geometry. In communication systems, a 'virtual array' may refer to clusters of distributed MIMO wireless communication nodes that form a virtual cooperative network to enhance the mobile link capacity such as presented in [Doh03]. In this thesis, virtual array refers to the resulting maximum baseline due to the arrangement of the physical antennas, which are closely spaced together.

It is not the intention of the work here to design an optimized physical nor virtual array, hence the simplest case, which is the uniform linear array (ULA) is considered here. Before the virtual arrays are explained, a few nomenclatures used in this chapter ought to be clarified as follows:

- P is the number of elements at the transmit array.
- Q is the number of elements at the receive array.
- $\vec{x}$ is the Cartesian coordinates position vector that can be decomposed to (x, y, z), with its direction relative to the origin at $\vec{x}_0 = (0, 0, 0)$.
- The p-th transmit antenna is located at $\vec{x}_{\mathrm{T}_p} \in \mathbb{R}^3$, where the upright subscript 'T' stands for 'transmitter' and p is the index of the transmit antenna, where $p = 0, 1, 2, ..., P-1$.
- The q-th receive antenna is located at $\vec{x}_{\mathrm{R}_q} \in \mathbb{R}^3$, where the upright subscript 'R' stands for 'receiver' and q is the index of the receive antenna with $q = 0, 1, 2, ..., Q-1$.
- The center of the antenna array, which is also taken as the origin of the local coordinate system is defined as $\vec{x}_0 = (0, 0, 0)$.

3.1 Virtual array concept

Consider firstly a linear antenna array consisting of P transmit elements, all spaced equally. This is known as a uniform linear array (ULA). If the P antennas are transmitting the same waveform but with different complex weighting factors at each antenna, the resulting radiation pattern can be shaped and focused to look a certain direction. By changing the weighting factors in terms of the feed current amplitude and phase, the radiation pattern and direction of the main beam can also be changed as dictated by the antenna array theory. This is frequently called the 'conventional array' and is the basis of the phased-array radar systems.

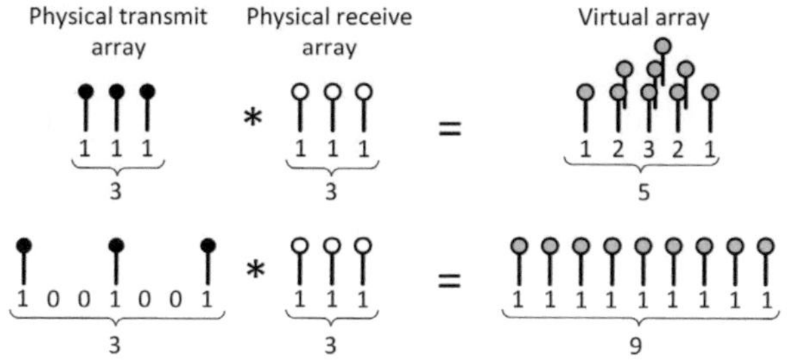

(a) 1D linear array.

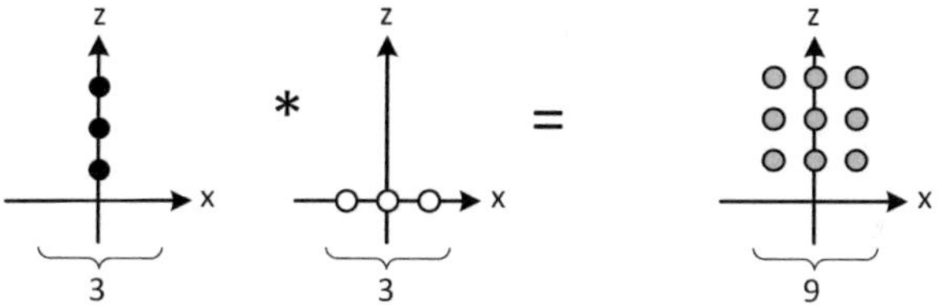

(b) 2D linear array.

Fig. 3.1: Virtual antenna array theory for a 1D and 2D linear array.

Now consider the same ULA, this time with each antenna transmitting an independent or orthogonal waveform. Due to this independence between the antenna elements, the radiated waveforms do not add up to result in a focusing of the main beam. Consider also that the Rx side consists of the same ULA with $Q = P$ elements as used at the Tx. In the radar channel, the waveforms are transmitted from the Tx to the targets and then backscattered to the Rx. At the Rx, all backscattered signals from the P transmitters are extracted for radar and DOA processing. The higher the number of P and Q, the more the Rx can 'focus in' on the backscattered signals from targets with significant RCS. 'Focusing in' on the target basically means detecting the target's peak based on scanning over the $P \times Q$ receive signals and maximizing the DOA image bins, which contain the relative angle between all qp pairs that most coincide with the target's angular direction. Hence the focusing of the transmit array's virtual main beam is actually done digitally at the Rx.

In order to achieve the maximum number of non-repetitive or non-redundant Rx signals, the physical transmit and receive linear arrays must be arranged to result in a maximum antenna baseline. The antenna array geometry can be visualized as a Dirac function, representing the antenna array elements' locations in space. The resulting virtual array is the convolution of the transmit and receive Dirac functions given by,

$$h_{\mathrm{T}}(\vec{x}) = \sum_{p=0}^{P-1} \delta(\vec{x} - \vec{x}_{\mathrm{T}_p}) \quad \text{and} \quad h_{\mathrm{R}}(\vec{x}) = \sum_{q=0}^{Q-1} \delta(\vec{x} - \vec{x}_{\mathrm{R}_q}). \tag{3.1}$$

Thus, the function to characterize the antenna element location in the virtual array is

$$h_{\mathrm{vir}} = (h_{\mathrm{T}} * h_{\mathrm{R}})(\vec{x}), \tag{3.2}$$

where $[*]$ is the convolution operation. Consider an example, where the transmit and receive arrays consist of three elements each with each element spaced $d = \lambda_{\mathrm{c}}/2$. The element spacing d is chosen according to the ULA theory to give no grating lobes within the $-90°$ to $90°$ FOV angle. An array with the notation $\{1 \quad 1 \quad 1\}$ is a contiguous or filled array, meaning that there is one antenna (at the position of '1') spaced d apart from the next antenna. If both the Tx and Rx use the same ULA, the resulting virtual array would be

$$\left\{1 \quad 1 \quad 1\right\} * \left\{1 \quad 1 \quad 1\right\} = \underbrace{\left\{1 \quad 2 \quad 3 \quad 2 \quad 1\right\}}_{\text{5 elements}}. \tag{3.3}$$

The resulting virtual array has five non-redundant elements and four repeated elements as depicted in Fig. 3.1(a)(top). It means that some virtual antennas are over-represented. This can also be seen from the antennas' channel response matrix $\mathbf{H}_{\mathrm{vir}}$

$$\mathbf{H}_{\text{vir}} \propto \mathbf{H}_{\text{T}} \otimes \mathbf{H}_{\text{R}} = \begin{pmatrix} e^{j\kappa\,-1d} \\ e^{j\kappa\,0d} \\ e^{j\kappa\,+1d} \end{pmatrix} \begin{pmatrix} e^{j\kappa\,-1d} & e^{j\kappa\,0d} & e^{j\kappa\,+1d} \end{pmatrix}$$
$$= \begin{pmatrix} e^{j\kappa\,-2d} & e^{j\kappa\,-1d} & e^{j\kappa\,0d} \\ e^{j\kappa\,-1d} & e^{j\kappa\,0d} & e^{j\kappa\,+1d} \\ e^{j\kappa\,0d} & e^{j\kappa\,+1d} & e^{j\kappa\,+2d} \end{pmatrix}, \tag{3.4}$$

where [⊗] is the outer product (for 1D matrices), a subset of the Kronecker product (for 2D matrices) and $\kappa = \beta \cos\vartheta = \frac{2\pi}{\lambda_c} \cos\vartheta$, with ϑ being the angle of far-field transmit waves shown in Fig. 3.2. It can be seen that there are repetitive matrix elements with the same phase rotation, resulting only in five unique responses and four repeated terms. Substituting either the transmit or receive array with a sparse array, the Dirac response becomes

$$\begin{Bmatrix}1 & 0 & 0 & 1 & 0 & 0 & 1\end{Bmatrix} * \begin{Bmatrix}1 & 1 & 1\end{Bmatrix} = \underbrace{\begin{Bmatrix}1 & 1 & 1 & 1 & 1 & 1 & 1 & 1 & 1\end{Bmatrix}}_{\text{9 elements}}. \tag{3.5}$$

This is also reflected in the channel response matrix given in Eq. (3.6), where it can be seen that there are no repetitive channel coefficients, hence ensuring no redundancy of information.

$$\mathbf{H}_{\text{vir}} \propto \mathbf{H}_{\text{T}} \otimes \mathbf{H}_{\text{R}} = \begin{pmatrix} e^{j\kappa\,-3d} \\ e^{j\kappa\,0d} \\ e^{j\kappa\,+3d} \end{pmatrix} \begin{pmatrix} e^{j\kappa\,-1d} & e^{j\kappa\,0d} & e^{j\kappa\,+1d} \end{pmatrix}$$
$$= \begin{pmatrix} e^{j\kappa\,-4d} & e^{j\kappa\,-3d} & e^{j\kappa\,-2d} \\ e^{j\kappa\,-1d} & e^{j\kappa\,0d} & e^{j\kappa\,+1d} \\ e^{j\kappa\,+2d} & e^{j\kappa\,+3d} & e^{j\kappa\,+4d} \end{pmatrix} \tag{3.6}$$

The locations of the virtual array can be derived from the observation of Eq. (3.6), leading to

$$\begin{aligned}
e^{j\beta d_{\mathrm{vir}}} &= e^{j\beta d_{\mathrm{T}_p}} \cdot e^{j\beta d_{\mathrm{R}_q}} \\
\Rightarrow d_{\mathrm{vir}} &= d_{\mathrm{T}_p} + d_{\mathrm{R}_q} \\
\therefore \vec{x}_{\mathrm{vir}} &= \vec{x}_{\mathrm{T}_p} + \vec{x}_{\mathrm{R}_q}
\end{aligned} \tag{3.7}$$

where d_{T_p} and d_{R_q} correspond to the transmit and receive element spacings respectively. In one dimension, the element spacing of the virtual array d_{vir} is simply the sum of the transmit and receive array element spacings. This concept can be extended to a 2D virtual array simply by taking into account their coordinate locations as will be presented in the later sections.

Now that it is qualitatively known that the virtual antennas' positions are determined by the physical antennas' positions, the exact locations of these antennas will be derived in the following section. While it is not shown here, the processing for the 2D linear array is the same as for the 1D array except that one of the arrays must be aligned to the elevation plane while the other to the azimuth plane as depicted in Fig. 3.1(b). Consequently the maximum baseline in one direction (azimuth and elevation) is a factor of P smaller than the 1D array and will decrease the angular resolution of the 2D array. The comparison between the angular resolution of the 1D and 2D arrays will be presented at the end of this chapter.

3.2 Physical antenna array geometry

According to the antenna array theory [Bal05, ch. 6], the resulting radiation pattern of an antenna array is dependent on the radiation pattern of the single element pattern multiplied by the array factor (AF). While it is true that the MIMO antenna array is technically not a conventional array, nevertheless the AF method can be used as a quantitative measure since the digital beam focusing at the Rx is in fact the same as overlaying the radiation patterns of the transmit and receive arrays. In this section, only the AF is considered and a brief derivation of a 1D transmit-receive array is presented, which leads to the analysis of the physical antenna arrangement to result in a no-redundancy virtual array. This result is applicable to both the 1D and 2D ULA.

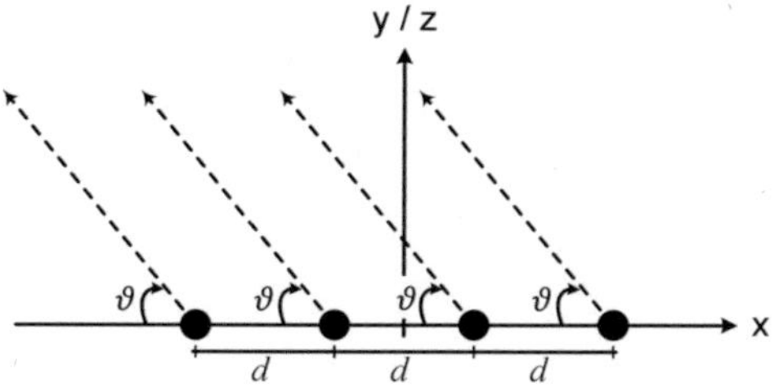

Fig. 3.2: Far-field observation of an array positioned along the x-axis. The radiation direction can be along the y- or z-axis.

Consider the AF for a P-element 1D ULA with each antenna element transmitting an independent signal. The phase ϑ of all incident rays to or from the array elements (assuming far-field conditions) as depicted in Fig. 3.2 is the same, therefore the AF can be simplified to,

$$\mathrm{AF} = 1 + \mathrm{e}^{\mathrm{j}\psi} + \mathrm{e}^{\mathrm{j}2\psi} + \cdots + \mathrm{e}^{\mathrm{j}(P-1)\psi}, \quad \psi = \beta d \cos\vartheta. \tag{3.8}$$

Multiplying both sides by $\mathrm{e}^{\mathrm{j}\psi}$ will result in

$$(\mathrm{AF})\,\mathrm{e}^{\mathrm{j}\psi} = \mathrm{e}^{\mathrm{j}\psi} + \mathrm{e}^{\mathrm{j}2\psi} + \cdots + \mathrm{e}^{\mathrm{j}(P-1)\psi} + \mathrm{e}^{\mathrm{j}P\psi}. \tag{3.9}$$

Subtracting Eq. (3.8) from Eq. (3.9) to obtain a bounded equation form,

$$\begin{aligned}
\mathrm{AF}\left(\mathrm{e}^{\mathrm{j}\psi} - 1\right) &= \left(-1 + \mathrm{e}^{\mathrm{j}P\psi}\right) \\
\mathrm{AF} &= \left[\frac{\mathrm{e}^{\mathrm{j}P\psi} - 1}{\mathrm{e}^{\mathrm{j}\psi} - 1}\right] \\
&= \mathrm{e}^{\mathrm{j}\frac{P-1}{2}\psi}\left[\frac{\mathrm{e}^{\mathrm{j}\frac{P}{2}\psi} - \mathrm{e}^{-\mathrm{j}\frac{P}{2}\psi}}{\mathrm{e}^{\mathrm{j}\frac{1}{2}\psi} - \mathrm{e}^{-\mathrm{j}\frac{1}{2}\psi}}\right] \\
&= \mathrm{e}^{\mathrm{j}\frac{P-1}{2}\psi}\left[\frac{\sin\left(\frac{P}{2}\psi\right)}{\sin\left(\frac{1}{2}\psi\right)}\right].
\end{aligned} \tag{3.10}$$

Making the origin of the Cartesian space as the center of the antenna array will shift the phase origin of the array to $\frac{P-1}{2}$ incurring a phase shift of $\mathrm{e}^{-\mathrm{j}\frac{P-1}{2}\psi}$, which simplifies Eq. (3.10) to,

$$\mathrm{AF} = \left[\frac{\sin\left(\frac{P}{2}\psi\right)}{\sin\left(\frac{1}{2}\psi\right)}\right]. \tag{3.11}$$

To find nulls of the array set AF = 0,

$$\begin{aligned}
\sin\left(\frac{P}{2}\beta d\cos\vartheta_{\mathrm{min}}\right) &= 0 \\
\Rightarrow \left(\frac{P}{2}\beta d\cos\vartheta_{\mathrm{min}}\right) &= \pm k\pi \\
\therefore \vartheta_{\mathrm{min}} &= \cos^{-1}\left[\frac{\lambda_{\mathrm{c}}}{2\pi d}\left(\pm\frac{2\pi k}{P}\right)\right] \\
\vartheta_{\mathrm{min}} &= \cos^{-1}\left[\pm k\frac{\lambda_{\mathrm{c}}}{dP}\right], \qquad \begin{array}{l} k = 1,2,3,\ldots \\ k \neq 0, P, 2P, \ldots \end{array}
\end{aligned} \tag{3.12}$$

To find the maxima of the array, the denominator of the AF is set to zero and AF is then reduced to a $\frac{\sin 0}{0}$ form. Hence when the value $k = \{0, P, 2P, \ldots\}$ is substituted into Eq. (3.12), the maxima will occur as given in Eq. (3.13). The first maximum is the main lobe that occurs at $k = 0$. The subsequent maxima are the grating lobes.

$$\vartheta_{\mathrm{max}} = \cos^{-1}\left[\pm k\frac{\lambda_{\mathrm{c}}}{dP}\right], \qquad k = 0, P, 2P, \ldots \tag{3.13}$$

The half-power beamwidth (HPBW) also called the 3 dB beamwidth occurs at the point when $\mathrm{AF} = \frac{1}{\sqrt{2}}$. The HPBW determines how sharp the main lobe of the antenna's radiation pattern is and is directly related to the angular resolution.

This will happen when $x = \pm 1.391$ for a $\frac{\sin x}{x} = \frac{1}{\sqrt{2}}$ function [9]. Therefore the angle at the 3 dB point from the main lobe maximum is,

$$\begin{aligned} \frac{P}{2}\beta d \cos\vartheta_{3\,\mathrm{dB}} &= \pm 1.391 \\ \vartheta_{3\,\mathrm{dB}} &= \cos^{-1}\left[\frac{\lambda_\mathrm{c}}{2\pi d}\left(\pm\frac{2.782}{P}\right)\right] \\ \vartheta_{3\,\mathrm{dB}} &= \frac{\pi}{2} - \sin^{-1}\left[\frac{\lambda_\mathrm{c}}{2\pi d}\left(\pm\frac{2.782}{P}\right)\right] \\ \vartheta_{3\,\mathrm{dB}} &\approx \left[\frac{\pi}{2} - \frac{\lambda_\mathrm{c}}{2\pi d}\left(\pm\frac{2.782}{P}\right)\right], \quad \text{if} \quad d \gg \lambda_\mathrm{c} \end{aligned} \tag{3.14}$$

The HPBW for a symmetrical main lobe is then,

$$\vartheta_{\mathrm{HPBW}} = 2\,|\vartheta_{\max} - \vartheta_{3\,\mathrm{dB}}| \qquad [\mathrm{rad}]. \tag{3.15}$$

Since there is no phase progression between the input currents of the antenna elements and since the center of the antenna is at the origin, this means that the main lobe is centered around 0°. Substituting $k = 0$ in Eq. (3.13) and putting it along with Eq. (3.14) into Eq. (3.15), the HPBW can be simplified to

$$\vartheta_{\mathrm{HPBW}} \approx \frac{2.782\,\lambda_\mathrm{c}}{\pi P d_\mathrm{T}} \qquad [\mathrm{rad}], \tag{3.16}$$

with d substituted with d_T to keep with the nomenclature for the transmit array.

3.2.1 Location of the transmit and receive antennas

Let the transmit array be the sparse array, while the Rx array is the filled array. In order to overlay both their radiation patterns to result in one main beam, the maxima of transmit array $\vartheta_{\max,\mathrm{T}}$ is made to coincide with minima of Rx array $\vartheta_{\min,\mathrm{R}}$ except at the main lobe. In accordance with Eq. (3.12) and Eq. (3.13), these are given by

$$\begin{aligned} \vartheta_{\max,\mathrm{T}} &= \cos^{-k}\left[\pm k\frac{\lambda_\mathrm{c}}{P d_\mathrm{T}}\right], \qquad k = P, 2P, ... \\ &= \cos^{-1}\left[\pm l\frac{\lambda_\mathrm{c}}{d_\mathrm{T}}\right] \qquad l = \tfrac{k}{P} = 1, 2, 3, ... \end{aligned} \tag{3.17}$$

9 Refer to $\frac{\sin x}{x}$ chart in Appendix I in [Bal05]

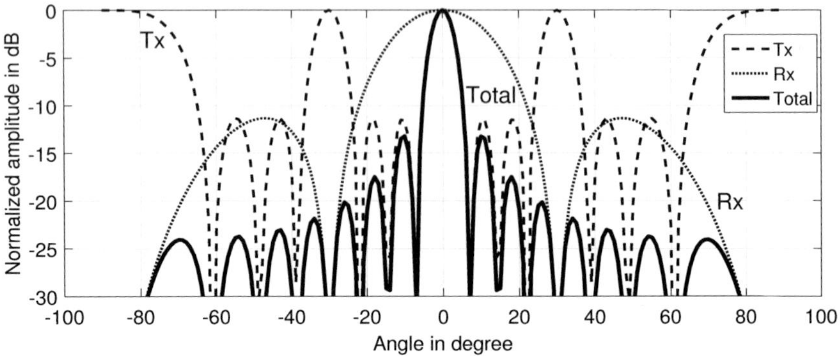

Fig. 3.3: MIMO equivalent array factor radiation pattern for 4×4 1D linear array configuration with $d_{\mathrm{R}} = \frac{\lambda_{\mathrm{c}}}{2}$ and $d_{\mathrm{T}} = 4 \cdot d_{\mathrm{R}}$.

and

$$\vartheta_{\mathrm{min,R}} = \cos^{-1}\left[\pm k \frac{\lambda_{\mathrm{c}}}{d_{\mathrm{R}} Q}\right], \qquad \begin{array}{l} k = 1,2,3,... \\ k \neq 0, Q, 2Q... \end{array} \tag{3.18}$$

Equating the null indices l in Eq. (3.17) and the maximum indices k in Eq. (3.18) will yield

$$\begin{aligned} \vartheta_{\mathrm{max,T}} &= \vartheta_{\mathrm{min,R}}, \qquad k = l \\ l \frac{\lambda_{\mathrm{c}}}{d_{\mathrm{T}}} &= k \frac{\lambda_{\mathrm{c}}}{d_{\mathrm{R}} Q} \\ \therefore d_{\mathrm{T}} &= Q d_{\mathrm{R}}, \end{aligned} \tag{3.19}$$

where the result is equivalent to the findings shown in Eq. (3.5) and Fig. 3.1(a). It can be seen that the sparse array's element spacing (the transmit array) is dependent on the number of elements available in the filled array and the filled array's element spacing. For illustration, let us consider a standard antenna array element spacing of $d_{\mathrm{R}} = \lambda_{\mathrm{c}}/2$, where within the angles of $[-90°, 90°]$ there are no grating lobes, as shown in Fig. 3.3 (Rx: dotted line). This results in $d_{\mathrm{T}} = Q \cdot \lambda_{\mathrm{c}}/2$, implying grating lobes within the $[-90°, 90°]$ angles as shown in Fig. 3.3 (Tx: dashed line). Multiplying the 'Tx' and 'Rx' radiation patterns will yield the 'Total' virtual radiation pattern, which has a much more focused main beam. From Eq. (3.15) and Eq. (3.19) the resulting HPBW for this transmit-receive array configuration is

$$\vartheta_{\mathrm{HPBW,TR}} \approx \frac{2.782\,\lambda_{\mathrm{c}}}{\pi P d_{\mathrm{T}}} = \frac{2.782\,\lambda_{\mathrm{c}}}{\pi P Q d_{\mathrm{R}}} \quad [\mathrm{rad}]. \tag{3.20}$$

The sparse and filled arrays can be either the transmit or receive arrays with no consequence to the resulting MIMO radiation pattern. The exact arrangement and positions of each physical and virtual antenna element can be determined from Eq. (3.7) and Eq. (3.19).

3.3 Beamsteering vectors

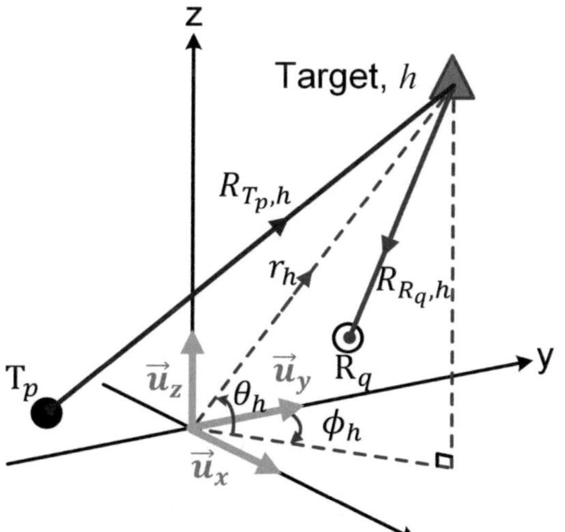

Fig. 3.4: Arbitrary transmit and receive antenna positions within the Cartesian and Spherical Coordinate Systems. The unit vectors $\vec{u}_x$, $\vec{u}_y$, $\vec{u}_z$ are as shown. If the target is located in the far-field of the antennas, $R_{\mathrm{T}_p,h}$ is assumed to be parallel to $R_{\mathrm{R}_q,h}$.

By normal conventions, a beamsteering vector is a vector comprising weights to steer and nullify the main beam at certain angle. For MIMO radars performing

DOA estimations, the beamsteering vector actually consists of counter-weights that equalizes the phase terms arising from the antenna array elements' positions. Every qp-th propagation path incurs an independent amount of phase shift relative to the center of the array, hence every path will require its own counter-weight. Since the qp-th path is the direct consequence of the physical location of the p-th and q-th antennas (as well as the targets), the counter-weight can be decomposed to only known variables, which are the locations of the antenna elements in the 3D Cartesian space. Before proceeding with the derivation of the beamsteering vector, a few assumptions need to be made, namely:

- The MIMO radar system is a quasi-monostatic one. This means that the Tx and Rx are located in the same platform, with closely-spaced antenna elements.
- The targets described are in the far-field region of the radar or antennas.
- The unit vectors from the origin $\vec{x}_0$ are defined with the vectors $\vec{u}_x$, $\vec{u}_y$, and $\vec{u}_z$ as shown in Fig. 3.4.
- In the case of the ULAs, d_T and d_R are the element spacing of the transmit and receive array respectively.

Following some of the steps done in Schuler [Sch07] and Nguyen [Ngu12], the derivation begins with the definition of the transmit and receive antennas and a single target within the 3D Cartesian space. The p-th transmit and q-th receive antenna elements are located arbitrarily at the coordinates $\vec{x}_{\mathrm{T}_p}, \vec{x}_{\mathrm{R}_q} \in \mathbb{R}^3$ respectively, while the h-th target relative to the origin $\vec{x}_0$ has the range vector $\vec{r}_h \in \mathbb{R}^3$. They are given by

$$\vec{x}_{\mathrm{T}_p} = \begin{pmatrix} x_{\mathrm{T}_p} \\ y_{\mathrm{T}_p} \\ z_{\mathrm{T}_p} \end{pmatrix} \qquad \vec{x}_{\mathrm{R}_q} = \begin{pmatrix} x_{\mathrm{R}_q} \\ y_{\mathrm{R}_q} \\ z_{\mathrm{R}_q} \end{pmatrix} \qquad \vec{r}_h = \begin{pmatrix} r_h \cos\theta_h \sin\phi_h \\ r_h \cos\theta_h \cos\phi_h \\ r_h \sin\theta_h \end{pmatrix}, \tag{3.21}$$

The geometry of one arbitrary transmit antenna, one arbitrary receive antenna and one target is shown in Fig. 3.4. The DOA of h-th object is estimated by a set of angular values (ϕ_h, θ_h), where ϕ_h is the azimuth angle and θ_h is the elevation angle. In order to find the counter weights of this particular propagation path, the total distance traveled from the the Tx to target and back to the Rx is first derived. To keep with the nomenclature of this chapter, the distance from the h-th target to the p-th Tx is $R_{\mathrm{T}_p,h}$, and this can be expressed as

$$\begin{aligned} R_{\mathrm{T}_p,h} &= \| \vec{r}_h - \vec{x}_{\mathrm{T}_p} \| = \left\| \begin{pmatrix} r_h \cos\theta_h \sin\phi_h \\ r_h \cos\theta_h \cos\phi_h \\ r_h \sin\theta_h \end{pmatrix} - \begin{pmatrix} x_{\mathrm{T}_p} \\ y_{\mathrm{T}_p} \\ z_{\mathrm{T}_p} \end{pmatrix} \right\| \\ &= \Big[r_h^2 - 2r_h x_{\mathrm{T}_p} \cos\theta_h \sin\phi_h \\ &\quad - 2r_h y_{\mathrm{T}_p} \cos\theta_h \cos\phi_h \\ &\quad - 2r_h z_{\mathrm{T}_p} \sin\theta_h + x_{\mathrm{T}_p}^2 + y_{\mathrm{T}_p}^2 + z_{\mathrm{T}_p}^2 \Big]^{1/2}. \end{aligned} \tag{3.22}$$

where $\|\cdot\|$ denotes the Euclidean norm of the vector. Using the binomial expansion theorem and taking into account that the Euclidean distance of the target r_h is much larger than the distance of the antenna array elements to the origin, $R_{\mathrm{T}_p,h}$ can be approximated as

$$R_{\mathrm{T}_p,h} \approx r_h - x_{\mathrm{T}_p} \cos\theta_h \sin\phi_h - y_{\mathrm{T}_p} \cos\theta_h \cos\phi_h - z_{\mathrm{T}_p} \sin\theta_h + \frac{x_{\mathrm{T}_p}^2 + y_{\mathrm{T}_p}^2 + z_{\mathrm{T}_p}^2}{2r_h}. \tag{3.23}$$

Defining the distance of the p-th transmit antenna element from the origin as

$$\vec{d}_{\mathrm{T}_p} = \vec{x}_{\mathrm{T}_p} - \vec{x}_0 = x_{\mathrm{T}_p} \vec{u}_x + y_{\mathrm{T}_p} \vec{u}_y + z_{\mathrm{T}_p} \vec{u}_z, \tag{3.24}$$

and in the same manner for q-th receive antenna element,

$$\vec{d}_{\mathrm{R}_q} = \vec{x}_{\mathrm{R}_q} - \vec{x}_0 = x_{\mathrm{R}_q} \vec{u}_x + y_{\mathrm{R}_q} \vec{u}_y + z_{\mathrm{R}_q} \vec{u}_z. \tag{3.25}$$

$R_{\mathrm{T}_p,h}$ can then be simplified further as

$$R_{\mathrm{T}_p,h} \approx r_h - x_{\mathrm{T}_p} \cos\theta_h \sin\phi_h - y_{\mathrm{T}_p} \cos\theta_h \cos\phi_h - z_{\mathrm{T}_p} \sin\theta_h + \frac{\|\vec{d}_{\mathrm{T}_p}\|^2}{2r_h}. \tag{3.26}$$

Using the same analysis, the distance between the h-th object and the q-th receiver can be obtained as

$$R_{\mathrm{R}_q,h} \approx r_h - x_{\mathrm{R}_q} \cos\theta_h \sin\phi_h - y_{\mathrm{R}_q} \cos\theta_h \cos\phi_h - z_{\mathrm{R}_q} \sin\theta_h + \frac{\|\vec{d}_{\mathrm{R}_q}\|^2}{2r_h}. \tag{3.27}$$

Therefore, the total traveling distance of the signal is

$$\begin{aligned} R_{\mathrm{T}_p,h} + R_{\mathrm{R}_q,h} \approx 2r_h &- (x_{\mathrm{T}_p} + x_{\mathrm{R}_q})\cos\theta_h \sin\phi_h \\ &- (y_{\mathrm{T}_p} + y_{\mathrm{R}_q})\cos\theta_h \cos\phi_h \\ &- (z_{\mathrm{T}_p} + z_{\mathrm{R}_q})\sin\theta_h + \frac{\|\vec{d}_{\mathrm{T}_p}\|^2 + \|\vec{d}_{\mathrm{R}_q}\|^2}{2r_h}. \end{aligned} \tag{3.28}$$

Based on Eq. (3.28), it can be seen that besides the desired range term of $2r_h$, there are also extra terms caused by the position of the transmit and receive array elements. In order to equalize the phase differences caused by the antenna element positions, a conjugate of the position must be taken for every pair of p and q antennas, whereby Eq. (3.28) would be equalized to leave only $2r_h$. Let this conjugate function K_h be defined as

$$\begin{aligned} K_h(\vec{d}_{\mathrm{T}_p}, \vec{d}_{\mathrm{R}_q}, r_0, \theta_0, \phi_0) = &- (x_{\mathrm{T}_p} + x_{\mathrm{R}_q})\cos\theta_0 \sin\phi_0 - (y_{\mathrm{T}_p} + y_{\mathrm{R}_q})\cos\theta_0 \cos\phi_0 \\ &- (z_{\mathrm{T}_p} + z_{\mathrm{R}_q})\sin\theta_0 + \frac{\|\vec{d}_{\mathrm{T}_p}\|^2 + \|\vec{d}_{\mathrm{R}_q}\|^2}{2r_0}, \end{aligned} \tag{3.29}$$

where θ_0 and ϕ_0 are the 'steering angles' of the elevation and azimuth directions respectively while r_0 is the actual distance from the h-th target to the origin. The angle compressed signal F_{AC} is simply the multiplication of the phase conjugate of K_h with the phase incurred by the distance traveled by the signal, expressed as

$$\begin{aligned} F_{\mathrm{AC}}(r_0, \theta_0, \phi_0) &= \sum_{q=0}^{Q-1} \sum_{p=0}^{P-1} \left[\mathrm{e}^{-\mathrm{j}2\beta K_h(\vec{d}_{\mathrm{T}_p}, \vec{d}_{\mathrm{R}_q}, r_0, \theta_0, \phi_0)} \right]^* \mathrm{e}^{-\mathrm{j}\beta\left(R_{\mathrm{T}_p,h} + R_{\mathrm{R}_q,h}\right)} \\ &= \underbrace{\mathrm{e}^{-\mathrm{j}\beta 2r_h}}_{\text{DOA term}} \cdot \sum_{q=0}^{Q-1} \sum_{p=0}^{P-1} \mathrm{e}^{\mathrm{j}\beta\left(x_{\mathrm{T}_p} + x_{\mathrm{R}_q}\right)(\cos\theta_h \sin\phi_h - \cos\theta_0 \sin\phi_0)} \\ &\qquad \mathrm{e}^{\mathrm{j}\beta\left(y_{\mathrm{T}_p} + y_{\mathrm{R}_q}\right)(\cos\theta_h \cos\phi_h - \cos\theta_0 \cos\phi_0)} \\ &\qquad \mathrm{e}^{\mathrm{j}\beta\left(z_{\mathrm{T}_p} + z_{\mathrm{R}_q}\right)(\sin\theta_h - \sin\theta_0)} \\ &\qquad \underbrace{\mathrm{e}^{\mathrm{j}\beta\left(|\vec{d}_{\mathrm{T}_p}|^2 + |\vec{d}_{\mathrm{R}_q}|^2\right)\left(\frac{1}{2r_h} - \frac{1}{2r_0}\right)}}_{\text{near-field term}}, \end{aligned} \tag{3.30}$$

with $[\cdot]^*$ denoting a matrix conjugate. The near-field term can be neglected since the object measurement is considered in the far-field only. This term contributes to a defocusing in the near-field [SS91] (see Fig. 7.11). Hence the angle compressed signal can be decomposed to only two relevant terms - the desired DOA term, and the phase terms arising from the geometry of the array elements, which are equalized by the counter weights. While $F_{\mathrm{AC}}(r_0, \theta_0, \phi_0)$ is indeed the desired DOA estimate, it is in an analytical form that does not help with the visualization of the DOA. It is hence of interest to modify the form of the phase conjugate function to result in steering vectors[10] that can be directly applied to the radar's receive matrix to form a DOA image[11]. From Eq. (3.30) it is clear that the relevant counter weights are given by

$$\mathbf{v}(\vec{x}_{\mathrm{T}_p}, \vec{x}_{\mathrm{R}_q}, \theta, \phi) = \left[\mathrm{e}^{\mathrm{j}\beta(x_{\mathrm{T}_p}+x_{\mathrm{R}_q})\cos\theta_0 \sin\phi_0} \cdot \mathrm{e}^{\mathrm{j}\beta(y_{\mathrm{T}_p}+y_{\mathrm{R}_q})\cos\theta_0 \cos\phi_0} \cdot \mathrm{e}^{\mathrm{j}\beta(z_{\mathrm{T}_p}+z_{\mathrm{R}_q})\sin\theta_0} \right]^*. \tag{3.31}$$

Separating the terms for the transmit and receive counter weights for easier implementation in the signal processing domain, this results in the transmit beamsteering vector, which is a 3D matrix is given by

$$\mathbf{v}_{\mathrm{T}}(\vec{x}_{\mathrm{T}_p}, \theta_0, \phi_0) = \begin{pmatrix} \mathrm{e}^{\left(\mathrm{j}\beta(x_{\mathrm{T}_1}\cos\theta_0\sin\phi_0 + y_{\mathrm{T}_1}\cos\theta_0\cos\phi_0 + z_{\mathrm{T}_1}\sin\theta_0)\right)} \\ \mathrm{e}^{\left(\mathrm{j}\beta(x_{\mathrm{T}_2}\cos\theta_0\sin\phi_0 + y_{\mathrm{T}_2}\cos\theta_0\cos\phi_0 + z_{\mathrm{T}_2}\sin\theta_0)\right)} \\ \vdots \\ \mathrm{e}^{\left(\mathrm{j}\beta(x_{\mathrm{T}_P}\cos\theta_0\sin\phi_0 + y_{\mathrm{T}_P}\cos\theta_0\cos\phi_0 + z_{\mathrm{T}_P}\sin\theta_0)\right)} \end{pmatrix}^* . \tag{3.32}$$

Analyzing the receive steering vector $\mathbf{v}_{\mathrm{R}}$ the same way yields

$$\mathbf{v}_{\mathrm{R}}(\vec{x}_{\mathrm{R}_q}, \theta_0, \phi_0) = \begin{pmatrix} \mathrm{e}^{\left(\mathrm{j}\beta(x_{\mathrm{R}_1}\cos\theta_0\sin\phi_0 + y_{\mathrm{R}_1}\cos\theta_0\cos\phi_0 + z_{\mathrm{R}_1}\sin\theta_0)\right)} \\ \mathrm{e}^{\left(\mathrm{j}\beta(x_{\mathrm{R}_2}\cos\theta_0\sin\phi_0 + y_{\mathrm{R}_2}\cos\theta_0\cos\phi_0 + z_{\mathrm{R}_2}\sin\theta_0)\right)} \\ \vdots \\ \mathrm{e}^{\left(\mathrm{j}\beta(x_{\mathrm{R}_Q}\cos\theta_0\sin\phi_0 + y_{\mathrm{R}_Q}\cos\theta_0\cos\phi_0 + z_{\mathrm{R}_Q}\sin\theta_0)\right)} \end{pmatrix}^* . \tag{3.33}$$

[10] a matrix that can be generated easily with digital signal processing software such as Matlab
[11] featuring a combination of range/Doppler with azimuth/elevation angles

The virtual beamsteering matrix is then expressed by,

$$\begin{aligned}&\mathbf{v}(\vec{x}_{\mathrm{T}_p},\vec{x}_{\mathrm{R}_q},\theta_0,\phi_0)\\&=\mathbf{v}_{\mathrm{R}}(\vec{x}_{\mathrm{R}_q}\theta_0,\phi_0)\otimes\mathbf{v}_{\mathrm{T}}(\vec{x}_{\mathrm{T}_p}\theta_0,\phi_0)\\&=\left(\mathbf{v}_{1,1}(\theta_0,\phi_0)\quad\cdots\quad\mathbf{v}_{1,P}(\theta_0,\phi_0)\quad\cdots\quad\mathbf{v}_{2,P}(\theta_0,\phi_0)\quad\cdots\quad\mathbf{v}_{Q,P}(\theta_0,\phi_0)\right)^{\mathsf{T}},\end{aligned}\tag{3.34}$$

where [⊗] here is the Kronecker product. Hence the virtual beamsteering matrix is a vector consisting only of the known physical transmit and receive array locations with the steering angles θ_0 and ϕ_0 sweeping over $\theta_0 \in [0,\pi]$ and $\phi_0 \in [-\pi/2,\pi/2]$.

3.3.1 1D antenna geometry beamsteering vector

A 1D ULA transmit and receive array is as shown in Fig. 3.5(a). The antenna array elements are positioned along the x-axis to result in the virtual array as shown in Fig. 3.5(c). A radar with such an antenna configuration will be able to estimate the x- and y- axes with the y-axis being the range and the x-axis being the cross-range (translated from the azimuth angle). Hence this is a 2D radar estimation. Since the antenna arrays are located only on the x-axis, elevation estimation is impossible. Setting $\theta_0 = 0°$ to result in $\cos\theta_0 = 1$, Eq. (3.31) reduces to

$$\mathbf{v}(\vec{x}_{\mathrm{T}_p},\vec{x}_{\mathrm{R}_q},\phi_0)=\left[\mathrm{e}^{\mathrm{j}\beta(x_{\mathrm{T}_p}+x_{\mathrm{R}_q})\sin\phi_0}\right]^*.\tag{3.35}$$

It is now of interest to see how accurately the steering angles coincide with the 'measuring angles' of ϕ_h, as well as the ambiguities that might arise due to the element spacing. With this simplest case of an equidistant 1D ULA, where all transmit and receive array elements are spaced d_{T} and d_{R} throughout respectively, taking only the phase terms due to the array geometry in Eq. (3.30) and modifying it by the conditions imposed in Eq. (3.35), the virtual AF is then

$$\mathrm{AF}_{\mathrm{vir,1D}}=\sum_{q=0}^{Q-1}\mathrm{e}^{\mathrm{j}\beta q d_{\mathrm{R}}(\sin\phi_h-\sin\phi_0)}\sum_{p=0}^{P-1}\mathrm{e}^{\mathrm{j}\beta p d_{\mathrm{T}}(\sin\phi_h-\sin\phi_0)}.\tag{3.36}$$

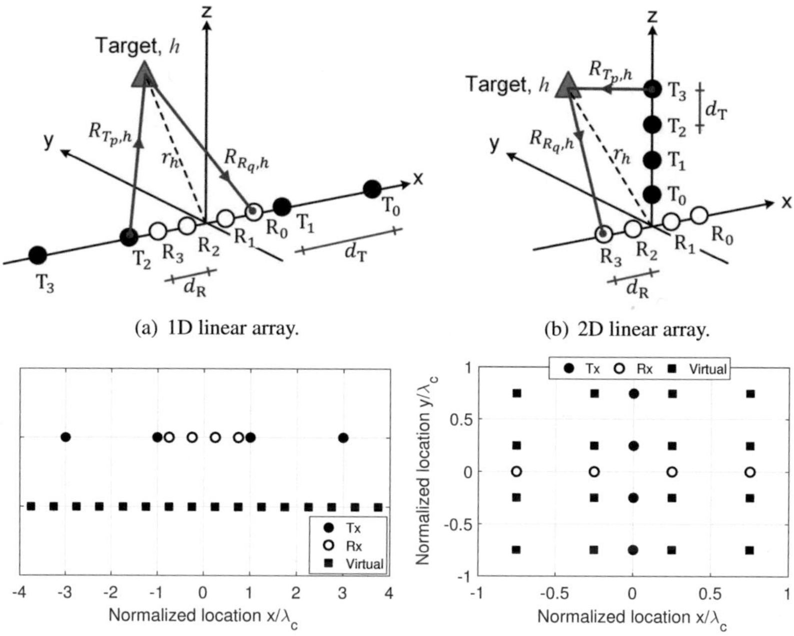

(a) 1D linear array.

(b) 2D linear array.

(c) Geometry of 1D physical and virtual antennas.

(d) Geometry of 2D physical and virtual antennas.

Fig. 3.5: Antenna geometry of linear 1D and 2D arrays located according to the Cartesian Coordinate System, with the y-axis being the propagation direction.

Using a similar mathematical manipulation as used in Eq. (3.10), $\mathrm{AF}_{\mathrm{vir,1D}}$ reduces to

$$\mathrm{AF}_{\mathrm{vir,1D}} = \frac{\sin\left(\frac{Q\beta}{2} d_{\mathrm{R}} (\sin\phi_h - \sin\phi_0)\right)}{\sin\left(\frac{\beta}{2} d_{\mathrm{R}} (\sin\phi_h - \sin\phi_0)\right)} \cdot \frac{\sin\left(\frac{P\beta}{2} d_{\mathrm{T}} (\sin\phi_h - \sin\phi_0)\right)}{\sin\left(\frac{\beta}{2} d_{\mathrm{T}} (\sin\phi_h - \sin\phi_0)\right)}. \quad (3.37)$$

For illustration, the settings $d_{\mathrm{R}} = \lambda_{\mathrm{c}}/2$ and $d_{\mathrm{T}} = Q d_{\mathrm{R}}$ for $P = Q = 4$ are used. Plotting both terms due to the transmit and receive arrays in Eq. (3.37) individually by sweeping over all combinations of $\phi_h - \phi_0$ values result in the plots in Fig. 3.6(a) and 3.6(b). When these two plots are multiplied, they yield the total virtual AF as shown in Fig. 3.6(c).

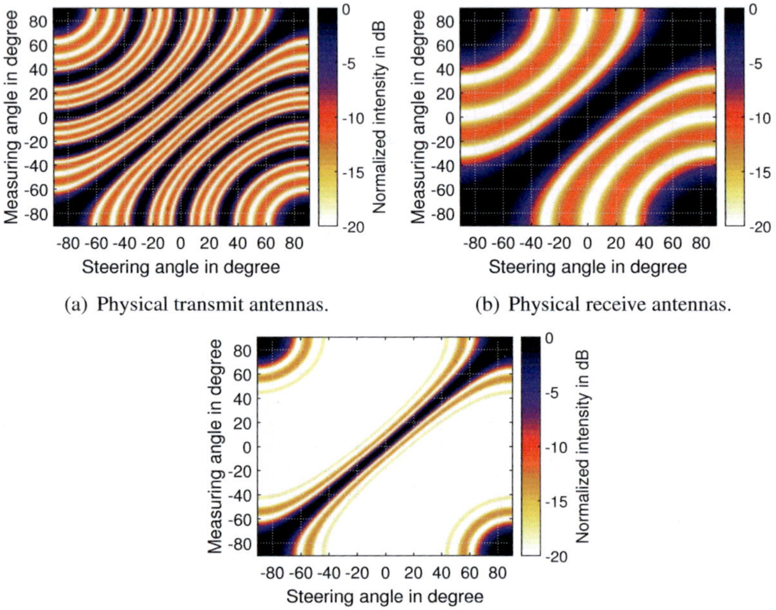

(a) Physical transmit antennas.

(b) Physical receive antennas.

(c) Resulting virtual antennas.

Fig. 3.6: Angular ambiguity plots of a 1D ULA of 4 × 4 transmit and receive uniform linear array antennas and resulting virtual antenna array.

While theoretically for such a configuration the ambiguities will only appear at ±90° due to grating lobes, it is evident here (for the 1D 4 × 4 case) that for the steering angles at $-70^\circ < \phi_0 < 70^\circ$, all steering angles coincide with all measuring angles as given by the diagonal line. At $\phi_0 = \pm 80^\circ$ however ambiguities start to appear due to the cyclic shift of the falling edge of the main beam (which appears at the corners of all plots in Fig. 3.6).

Shown in Fig. 3.7(a) is the cut of the angular ambiguity plot in Fig. 3.6(c) at $\phi_0 = 80^\circ$ for the 4 × 4 configuration. An object at this angle would result in two measuring angles of -80° and 80° due to the cyclic shift of the broad main lobe. This then results in the same maximum amplitude at -80° and 80°. To improve the unambiguous angular FOV, a higher number of transmit and receive antenna elements can be used to further 'sharpen' the resulting virtual main beam as shown in Fig. 3.7(b) for an 8×8 configuration. Here, the main lobe is focused at 80° creating no ambiguity.

Hence the angular ambiguity plot of the ULAs featuring the steering vs. measuring angles is a convenient tool to determine the angular FOV of the antenna arrays used since the ambiguity regions are not obvious in the MIMO equivalent radiation pattern plot in Fig. 3.3. Meanwhile the angular resolution can be gauged by the thickness of the diagonal line of Fig. 3.6(c). The usage of this tool and the consequence to the FOV of the radar's DOA estimation is presented in Chapter 7.

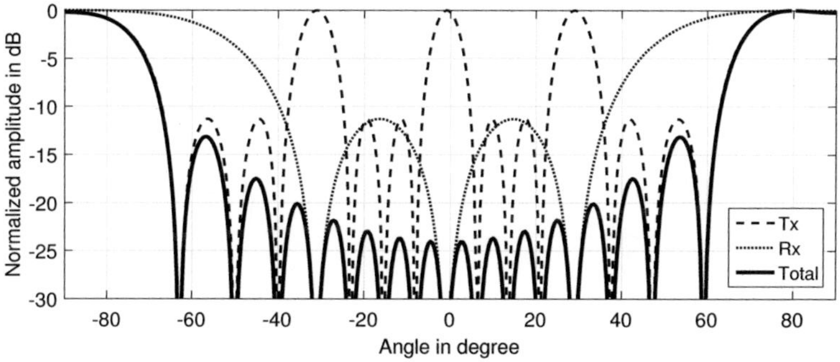

(a) 4×4 configuration at $\phi_0 = 80°$. Angular estimation results in $\phi_h = -80°$ and 80° creating ambiguity.

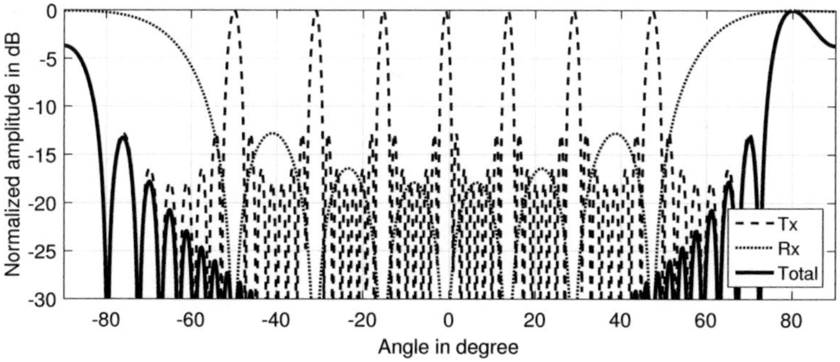

(b) 8×8 configuration at $\phi_0 = 80°$. No ambiguity at the measuring angle of $\phi_h = 80°$.

Fig. 3.7: A horizontal cut at $\phi_0 = 80°$ of the total ambiguity plot of Fig. 3.6(c) for a 1D ULA of 4×4 and 8×8 configuration.

3.3.2 2D antenna geometry beamforming vector

For a 3D DOA estimation, one of the ULAs is placed along the z-axis to illuminate the elevation direction while the other is placed along with x-axis as shown in Fig. 3.5(b). This antenna geometry will result in the virtual array geometry as shown in Fig. 3.5(d). Since the ULAs are located only on the x- and z-axes hence $\cos\phi_0 = 0°$ for the y-axis, reducing Eq. (3.31) to

$$\mathbf{v}(\vec{x}_{\mathrm{T}_p}, \vec{x}_{\mathrm{R}_q}, \theta_0, \phi_0) = \left[\mathrm{e}^{\mathrm{j}\beta(x_{\mathrm{T}_p}+x_{\mathrm{R}_q})\cos\theta_0\sin\phi_0} \mathrm{e}^{\mathrm{j}\beta(z_{\mathrm{T}_p}+z_{\mathrm{R}_q})\sin\theta_0} \right]^*. \tag{3.38}$$

Using the same analysis as the 1D ULA, the 2D virtual AF can be expressed as

$$\begin{aligned} \mathrm{AF}_{\mathrm{vir,2D}} &= \sum_{q=0}^{Q-1} \mathrm{e}^{\mathrm{j}\beta q d_{\mathrm{R}}(\cos\theta_h\sin\phi_h - \cos\theta_0\sin\phi_0)} \sum_{p=0}^{P-1} \mathrm{e}^{\mathrm{j}\beta p d_{\mathrm{T}}(\sin\theta_h - \sin\theta_0)} \\ &= \frac{\sin\left(\frac{Q\beta}{2} d_{\mathrm{R}}(\cos\theta_h\sin\phi_h - \cos\theta_0\sin\phi_0)\right)}{\sin\left(\frac{\beta}{2} d_{\mathrm{R}}(\cos\theta_h\sin\phi_h - \cos\theta_0\sin\phi_0)\right)} \cdot \frac{\sin\left(\frac{P\beta}{2} d_{\mathrm{T}}(\sin\theta_h - \sin\theta_0)\right)}{\sin\left(\frac{\beta}{2} d_{\mathrm{T}}(\sin\theta_h - \sin\theta_0)\right)}. \end{aligned} \tag{3.39}$$

With a 2D configuration, the 2D virtual AF radiation pattern can also be visualized. Here two configurations of the same array geometry are shown for comparison. Using the standard $d_{\mathrm{T}} = d_{\mathrm{R}} = \lambda_{\mathrm{c}}/2$ for both configurations, Fig. 3.8(a) shows the virtual array radiation pattern for the 4×4 configuration, while Fig. 3.8(c) shows the one for the 16×16 configuration.

In the same way as for the 1D ULA, the angular ambiguity plots are as depicted in Fig. 3.8(b) and 3.8(d) respectively. These plots only show the 1D steering versus measuring angle of either the azimuth or elevation plane. When using Eq. (3.39) for plotting the azimuth plane, all $\theta_h = \theta_0 = 0°$ while for the elevation plane $\phi_0 = \phi_h = 0°$. Both the transmit and receive arrays having the same element spacing but positioned at different planes can no longer take advantage of the beam focusing as done in the 1D array. It is obvious when comparing the 1D case in Fig. 3.6(c) and the 2D case in Fig. 3.8(b) for the same 4×4 ULA configuration, that the maximum baseline of the antennas has not been achieved for the 2D ULA case, hence the angular resolution is $P = 4$ times lower than for the 1D ULA case. This is then proven by comparing the 1D ULA's angular resolution in Fig. 3.6(c) to the 2D 16×16 ULA case in Fig. 3.8(d), where it can be seen that they are the same. For the 2D ULA case, the angular resolution in both the elevation and azimuth can also be gauged from the virtual array radiation pattern plots based on the size of the main beam at 0° for all angles.

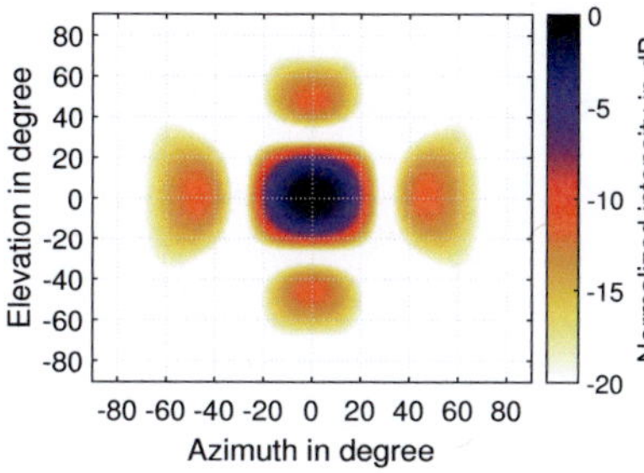

(a) 4×4: virtual array pattern.

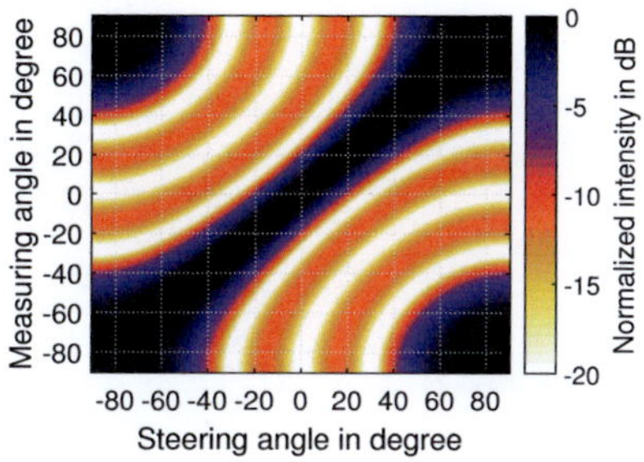

(b) 4×4: ambiguity plot of the azimuth or elevation plane.

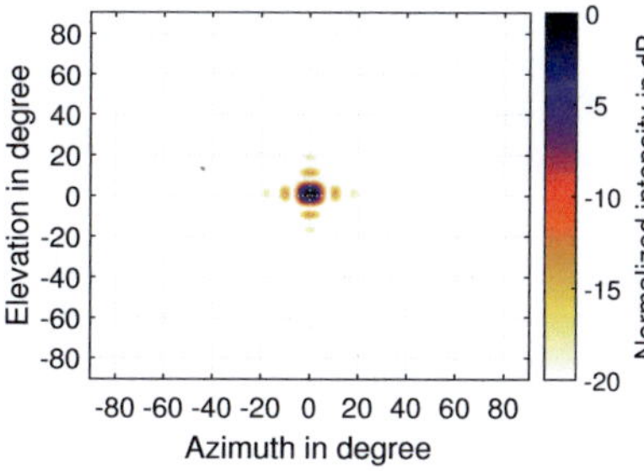

(c) 16×16: virtual array pattern.

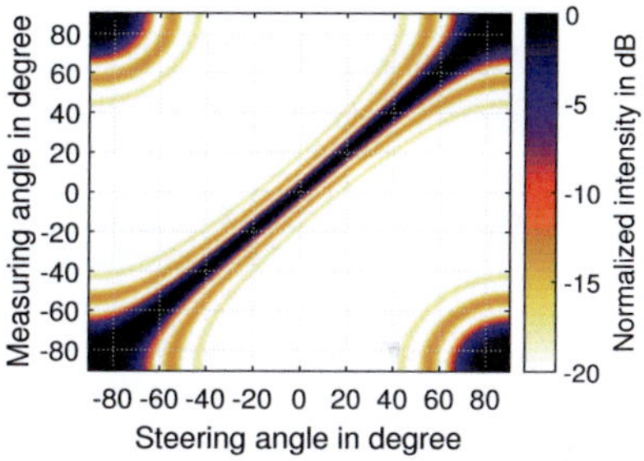

(d) 16×16: ambiguity plot of the azimuth or elevation plane.

Fig. 3.8: Resulting virtual antenna radiation pattern for a 2D ULA of 4×4 and 16×16 transmit and receive uniform linear array antennas. The angular ambiguity plots are the same for both the azimuth (with $\theta_0 = \theta_h = 0°$) and elevation planes (with $\phi_0 = \phi_h = 0°$).

3.4 Direction-of-Arrival estimation

The range and Doppler effect on the radar's receive signal is independent of the effect of the DOA term. Hence angular estimation can be done on the range and Doppler compressed signal $\mathbf{I}_{qp}(k,l)$, from Eq. (2.47) instead of on the raw received signal. Let the processed radar matrix from $P \times Q$ antennas be declared in a matrix as follows,

$$\mathbf{I}(p,q,k,l) = \begin{pmatrix} \mathbf{I}_{1,1}(k,l) & \cdots & \mathbf{I}_{1,P}(k,l) & \cdots & \cdots & \mathbf{I}_{2,P}(k,l) & \mathbf{I}_{Q,P}(k,l) \end{pmatrix}. \tag{3.40}$$

To apply the beamsteering vector from Eq. (3.34) to equalize $\mathbf{I}(p,q,k,l)$, a Frobenius product[12] is applied as follows,

$$\mathbf{FB}(k,l,\theta,\phi) = \langle \mathbf{I}(p,q,k,l), \mathbf{v}(\vec{x}_{\mathrm{T}_p}, \vec{x}_{\mathrm{R}_q}, \theta, \phi) \rangle_F \tag{3.41}$$

where $\mathbf{FB}(\theta,\phi)$ is the 4D DOA image matrix. This method is called the Fourier Beamforming method because the beamsteering vectors of the ULA can be applied on $\mathbf{I}(p,q,k,l)$ via a discrete Fourier transform (DFT). This has been successfully applied on an FMCW[13] radar with digital beamforming for DOA estimation in [HZZ11]. Due to the spectrally interleaved signal model used in this thesis however, the DFT method is not applicable here. A DFT over $\mathbf{I}(p,q,k,l)$ will result in the wrong equalization since the step size of the phase since the index k[14] from one $\mathbf{I}_{qp-1}(k,l)$ to the next $\mathbf{I}_{qp}(k,l)$ is not regular due to the influence of the user index u.

3.4.1 Angular resolution

The angular resolution is the minimum separation of two closely-spaced targets that must be adhered to for the radar to perform DOA estimation correctly. In this work, the angular resolution is always given in degrees equivalent to the virtual half power beamwidth $\vartheta_{\mathrm{HPBW,vir}}$ and it is fixed for a fixed antenna array configuration. The cross-range resolution is defined as the minimum separation between two targets that is dependent on their range from the radar. This can be visualized as shown in Fig. 3.9, where $\Delta r_{\mathrm{cross}}$ is the minimum separation (or cross-range resolution) between the two targets as stipulated by the virtual HPBW, $\vartheta_{\mathrm{HPBW,vir}}$. Applying Eq. (3.20) and using small angle approximation, the $\Delta r_{\mathrm{cross}}$ for a 1D array in meter can be written as

[12] sum of element-wise multiplication of vectors with $\langle A,B\rangle_F = \sum_{i,j}(a_{ij}b_{ij})$

[13] Frequency modulated continuous wave

[14] processed from the subcarrier index n_u

$$\begin{aligned}\Delta r_{\text{cross}} &= 2r\sin\frac{\vartheta_{\text{HPBW,vir}}}{2}\\ &\approx \frac{0.89\lambda_{\text{c}}}{d_{\text{T/R}}PQ}\cdot r\\ &\approx \frac{\lambda_{\text{c}}}{L_{\text{a}}}\cdot r \qquad [\text{m}].\end{aligned} \tag{3.42}$$

with $d_{\text{T/R}} = \min(d_{\text{T}}, d_{\text{R}})$, dependent on the arrangement of the transmit and receive antennas. The $\vartheta_{\text{HPBW,vir}}$ is oftentimes approximated with $\lambda_{\text{c}}/L_{\text{a}}$, where L_{a} is the effective length of the ULA. Although this is the theoretical resolution, practically one should take a factor of two to four times of this theoretical resolution for a good discernment of the targets since the value changes according to the physical antennas' radiation pattern and characteristics.

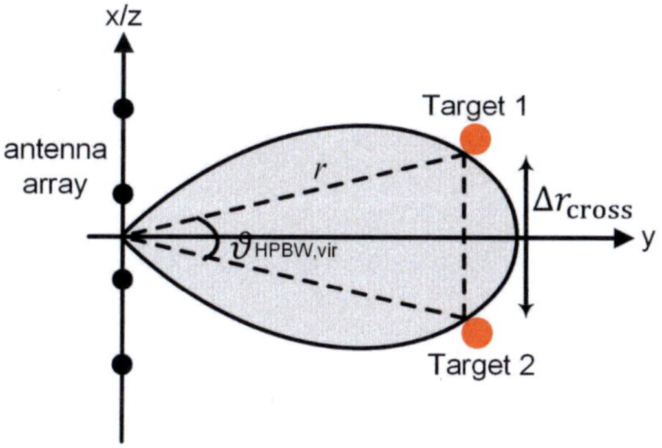

Fig. 3.9: Visualization of angular resolution and cross-range resolution.

Fig. 3.10 shows the dependency of the angular resolution on windowing, simulated with a 4×4 ULA configuration with $f_{\text{c}} = 24\,\text{GHz}$ and $d_{\text{R}} = \lambda_{\text{c}}/2$. The antennas were modeled as point sources, thus eliminating the influence of the element factor in the resulting virtual AF. Using Eq. (3.42), $\vartheta_{\text{HPBW,vir}} \approx 7.2°$, but the minimum possible simulated angle was around 10.6°. When no windowing is used, the two targets can be explicitly distinguished but at the expense of higher sidelobes as seen in the azimuth cut. When a Hamming window is applied, the sidelobes are suppressed but the targets are no longer distinguishable as separate targets.

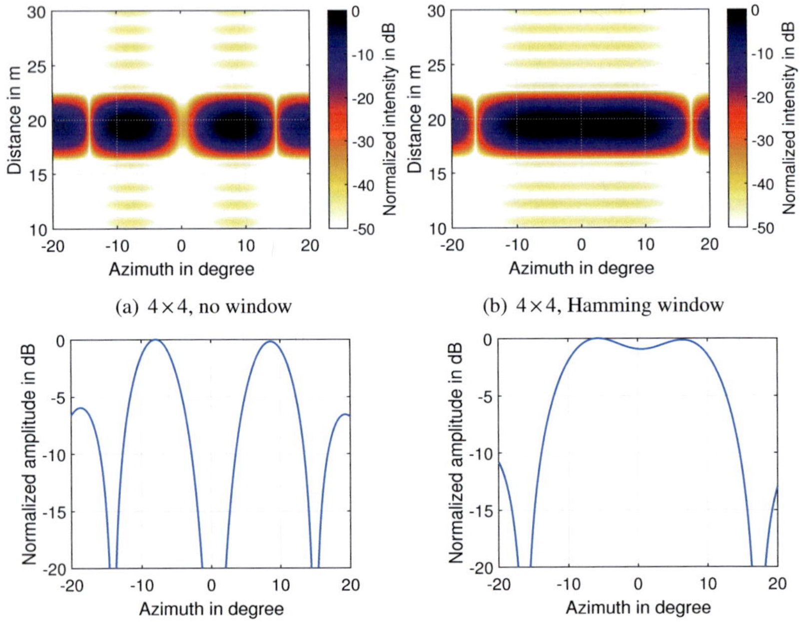

(a) 4×4, no window

(b) 4×4, Hamming window

(c) 4×4, no window cut

(d) 4×4, Hamming window cut

Fig. 3.10: Windowing effect on angular resolution with two targets at 10.6° apart using Fourier Beamforming method on a 4×4 1D array configuration.

The same simulation is now repeated with a higher number of antennas, where $P = Q = 8$ with the results shown in Fig. 3.11. With a larger effective aperture (hence a larger antenna baseline), the theoretical $\vartheta_{\mathrm{HPBW,vir}} \approx 1.8°$ but a minimum angle of 3.2° is required in the simulation. Likewise, the Hamming window distorts the DOA image rendering the two targets indistinguishable. Based on these results, a conclusion can be made, whereby the minimum angle of separation must always be larger than stipulated by the theoretical value especially when windowing is applied to the DOA image.

Some reserach has also shown that the physical antenna geometry and orientation will also affect the outcome of the angular resolution, based on the shape of the virtual AF radiation pattern, such as in [HMS+13]. Subspace methods like MUSIC [Sch86] and ESPRIT [PRK86] has also been successfully applied to obtain high angle accuracy with no sidelobes DOA image as presented in [SNSZ13]. This is however out of the scope of this thesis and will not be further discussed.

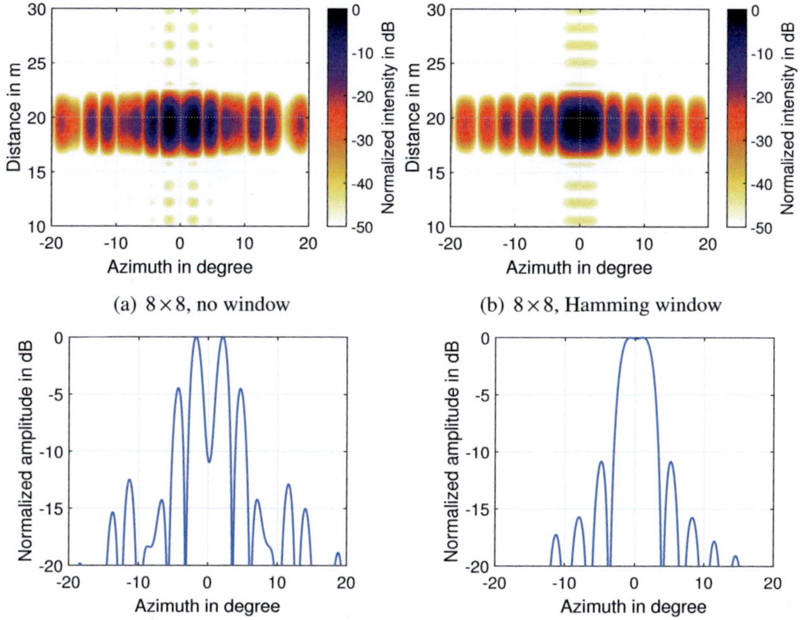

(a) 8×8, no window

(b) 8×8, Hamming window

(c) 8×8, no window cut

(d) 8×8, Hamming window cut

Fig. 3.11: Windowing effect on angular resolution with two targets at 3.2° apart using Fourier Beamforming method on a 8×8 1D array configuration.

3.5 Chapter 3 summary

The virtual array concept is used to form a maximum antenna baseline using a minimum number of physical antenna elements. This also results in the maximum number of non-redundant receive signals in terms of DOA.

The beamsteering vector is composed of coefficients or counter weights that equalize the effects of the antenna array elements' positions on the receive signal. From the beamsteering vector derivations, it can be seen that any type of antenna array can be used so long as they adhere to the no redundancy or minimum redundancy rule when it comes to the channel response.

The angular resolution is dependent on the number of antenna elements used (which in turn affects the $\vartheta_{\mathrm{HPBW,vir}}$) and the element spacing of d_{T} and d_{R}. Using $d_{\mathrm{R}} = \lambda_{\mathrm{c}}/2$ with $d_{\mathrm{T}} = Qd_{\mathrm{R}}$ will provide the theoretical maximum angular FOV of $\pm 70°$ due to the number of antennas used (see Fig. 3.6(c) and Fig. 3.8(d)). Real physical antenna arrays however are usually bulky and are unable to meet the requirement, which then leads to the narrowing of the angular FOV. The antenna characteristics are taken into account with the ambiguous regions demonstrated in the measurement verification in Chapter 7.

4 Communication and interference tolerance of the RadCom

This chapter aims to provide an insight about the communications portion of the RadCom. Since the goal of the MIMO RadCom is to function in a network, the ability to also exchange data signals between RadCom nodes becomes the major advantage of using a RadCom over other networked radar systems that are unable to communicate with each other via the same radar hardware. Here, it is of interest to see how a communication signal affects the radar signal (backscattered from objects), how the payload data recovery is made possible and the immunity of the radar when the signal from a communication partner turns into an interferer signal.

It is well known that OFDM signals are weak toward subcarrier misalignment caused by frequency offsets. While employing the spectrally interleaved signal model by assigning different transmit antennas with independent user indices can avoid intra- and inter-system (mutual) interference, channel (e.g. Doppler) and hardware imperfections (e.g. LO drift) usually make it impossible for the radars to function solely within their allotted set of frequency bands. Thus it is of interest to know how well the radar can tolerate interferers in terms of frequency offsets (δ) and signal-to-interferer power ratio at the radar's input ($\mathrm{SIR}_{\mathrm{in}}$) based on the quality metric of the radar's Signal-to-Noise plus Interferer at the output ($\mathrm{SNIR}_{\mathrm{out}}$).

4.1 Interference from another RadCom node

Imagine that there are multiple RadCom nodes in the vicinity. Whether these nodes are functioning cooperatively or not, there will be channel effects that will distort the orthogonality of the OFDM subcarriers. Even with the non-overlapping subcarrier assignment to all the nodes, due to the relative movements of the nodes with each other or the movement of the targets under surveillance, the orthogonality of the subcarriers will be compromised. These relative movements will incur a Doppler frequency, which is in essence, a frequency shift (or offset). Along with this, the RadCom nodes' hardware can possess an initial phase and/or frequency shift, which not only adds to the misalignment of the subcarriers, but also causes errors to the radar and DOA estimates. As such there are two major effects of interference from another RadCom node to consider namely, 1) the effect of the power of the interferer, and 2) the effect of the frequency shift.

Henceforth, the ego RadCom node doing the observation and estimation of the 3D+velocity parameters of its FOV is termed as the '**Radar**' or '**Radar node**', while the communication partner node of the same RadCom system is termed interchangeably as the '**Comm**' or '**Interferer**'.

4.1.1 Comm signal power

Let us assume here that there are two RadCom nodes in the vicinity with the distance of r_{comm} between them. One of them is the Radar and the other is the Comm. The Radar is making a measurement of the H targets that are r_h away respectively. The Comm on the other hand is sending a message to the Radar. Before proceeding with the effect of interference on the radar signal, it is useful to have an idea how much the Comm's signal power $P_{\mathrm{R,comm}}$, can overwhelm the desired backscattered signals from all H targets to the Radar $P_{\mathrm{R},h}$. For this basic demonstration it is sufficient to use the Friis equation to depict the power of Comm's signal at the Radar's receiver, and the radar equation for the desired backscattered signal power.

Here it is assumed that the Radar and the Comm transmit the same power P_{T} with identical transmit antenna gain G_{T} and receive antenna gain G_{R}. Assuming that the antennas of the Radar and Comm are pointed directly toward each other, the aforementioned powers can be expressed as

$$P_{\mathrm{R,comm}} = P_{\mathrm{T}} G_{\mathrm{T}} G_{\mathrm{R}} \cdot [L_{\mathrm{comm}}]^{-1} \qquad \text{with } L_{\mathrm{comm}} = \left(\frac{4\pi}{\lambda_{\mathrm{c}}}\right)^2 r_{\mathrm{comm}}^2$$
$$P_{\mathrm{R},h} = P_{\mathrm{T}} G_{\mathrm{T}} G_{\mathrm{R}} \cdot [L_h]^{-1} \qquad \text{with } L_h = \left(\frac{4\pi}{\lambda_{\mathrm{c}}}\right)^2 \sum_{h=0}^{H-1} \frac{4\pi r_h^4}{\sigma_h}, \tag{4.1}$$

where L_{comm} and L_h are the path loss terms and σ_h is the RCS of the h-th object. The signal-to-interference ratio at the input ($\mathrm{SIR_{in}}$) is then

$$\mathrm{SIR_{in}} = \frac{P_{\mathrm{R},h}}{P_{\mathrm{R,comm}}} = \frac{r_{\mathrm{comm}}^2}{4\pi} \sum_{h=0}^{H-1} \frac{\sigma_h}{r_h^4}. \tag{4.2}$$

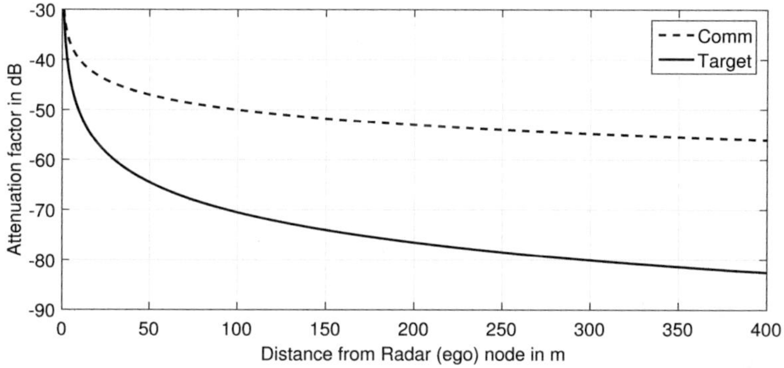

Fig. 4.1: Attenuation factor for communication signal vs. desired radar reflected signal from one target of $\sigma = 10\,\mathrm{m}^2$ over distance based on the Friis and radar equations.

To visualize how strong $P_{\mathrm{R,comm}}$ and $P_{\mathrm{R},h}$ are over difference distances, Fig. 4.1 shows the attenuation factor of the Comm's signal $[L_{\mathrm{comm}}]^{-1/2}$, and of the Radar's signal reflected from the target $[L_h]^{-1/2}$, for one target of RCS $\sigma = 10\,\mathrm{m}^2$ (approximately equivalent to a medium sized car at 24 GHz), which are directly proportional to the received powers. It can be seen that the Comm's signal attenuation is always lower than the Radar's for the same given Comm node and target distance. The reality is, should there be a a subcarrier misalignment, the Comm's signal power will leak into the Radar's allocated subchannels and overwhelm the weak targets' reflected signals. Depending on the $\mathrm{SIR_{in}}$, this will raise the noise floor and consequently lowers the $\mathrm{SNIR}_{\mathrm{out}}$ (that is inclusive of the processing gain G_{per}), and in the worst case scenario drowns the Radar's estimation matrix with a white noise-like characteristics.

4.1.2 Time delay due to distance

To built the mathematical expression for the Comm signal, the time delay experienced by the signal due to its distance to the Radar must also be considered. Analogous to Eq. (2.26) the phase rotation terms due to the range can be written as

$$\begin{aligned} \kappa_{r_w}(n_u) &= \mathrm{e}^{-\mathrm{j}2\pi(u+iN_{\mathrm{ch}})\Delta f \tau_h}, & n_u &= u + iN_{\mathrm{ch}} \\ &= \mathrm{e}^{-\mathrm{j}2\pi u \Delta f \tau_w} \cdot \mathrm{e}^{-\mathrm{j}2\pi(iN_{\mathrm{ch}})\Delta f \tau_w}, & i &= 0,1,...,\frac{N}{N_{\mathrm{ch}}} - 1 \\ &= \mathrm{e}^{\mathrm{j}\varphi_{\mathrm{R}_w}(u)} \cdot \mathrm{e}^{-\mathrm{j}2\pi i N_{\mathrm{ch}} \Delta f \tau_w}. \end{aligned} \tag{4.3}$$

with the subscript 'w' denoting the Comm and $\tau_w = \frac{r_w}{c_0}$ is the time delay of the Comm signal.

4.1.3 Carrier frequency offset

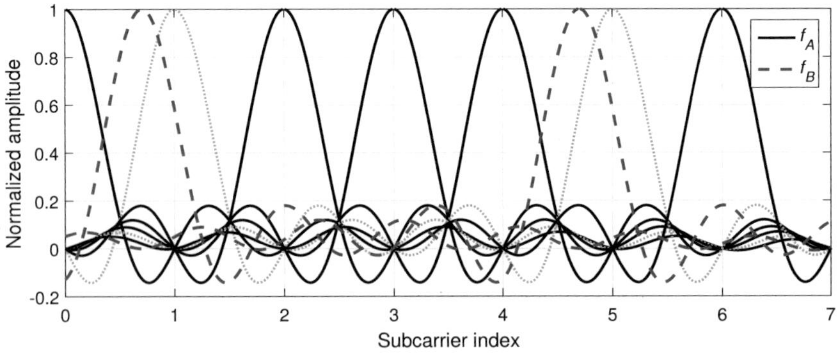

Fig. 4.2: Effect of frequency offset on OFDM subcarrier orthogonality. Signal f_B with $\delta_w T_0 = -0.3$ (dashed lines) is shifted onto signal f_A. The peaks of f_A no longer correspond to the zero-crossings of f_B. The faint dotted lines traces the location that signal f_B should be for orthogonality with signal f_A.

One of the effects of hardware imperfection is the carrier frequency offset (CFO). CFO is the frequency difference between the Comm's LO and the Radar's LO due to the phase-locked loop (PLL) locking onto different time and frequency reference sources. This cannot be prevented and can only be calibrated. The effect of the CFO $f_{\mathrm{LO},w}$ is the same as Doppler $f_{\mathrm{D},w}$ hence the frequency shift term can be combined using Eq. (2.29) to yield

$$\begin{aligned}\kappa_{\mathrm{D}_w}(n_u, m) &= \mathrm{e}^{\mathrm{j}2\pi\delta_w n_u \frac{T_0}{N}} \cdot \mathrm{e}^{\mathrm{j}2\pi\delta_w mT} \cdot \mathrm{e}^{\mathrm{j}2\pi\delta_w M_{\mathrm{train},w} T} \\ &= \mathrm{e}^{\mathrm{j}\varphi_{\mathrm{D}_w}(n_u)} \cdot \mathrm{e}^{\mathrm{j}2\pi\delta_w mT} \cdot \mathrm{e}^{\mathrm{j}\varphi_{\mathrm{train},w}}, \qquad \delta_w = f_{\mathrm{LO},w} + f_{\mathrm{D},w}.\end{aligned} \tag{4.4}$$

The frequency offset between the Radar and Comm can be visualized as shown in Fig. 4.2, where f_A is the Radar's signal and f_B is the Comm's signal with a frequency offset of $\delta_w T_0 = -0.3$. The loss of orthogonality is caused due to the peaks of the subcarriers of f_A no longer corresponding to the zero-crossings of f_B.

Effect of frequency offset on the payload data demodulation

The effect of a frequency offset on the orthogonality of the OFDM signal will be insightful and is discussed here. The nomenclature is taken from Fig. 2.2 with the utilization of a SISO radar model where all subcarriers are utilized for simplicity. Following Eq. (2.53), the recovered (or estimated) modulation symbol $\tilde{a}_m(n')$ will be

$$\tilde{a}_m(n') = \mathrm{e}^{\mathrm{j}(\Psi+\zeta)}\,\mathrm{e}^{\mathrm{j}2\pi\delta_w mT} \underbrace{\sum_{n=0}^{N-1} \mathrm{e}^{\mathrm{j}\frac{2\pi}{N}nk'}}_{\text{IDFT term}} a_m(n)\,\mathrm{e}^{-\mathrm{j}2\pi n\Delta f\tau} \underbrace{\sum_{\ell=0}^{N-1} \mathrm{e}^{-\mathrm{j}\frac{2\pi}{N}n'\ell} \cdot \mathrm{e}^{\mathrm{j}\frac{2\pi}{N}\ell\delta_w T_0}}_{\text{DFT term}}. \tag{4.5}$$

To isolate the effect of the CP ambiguity and frequency offset, assume here that the CP length is removed correctly hence $k' = \ell$. This will not affect the frequency offset. Grouping terms that are dependent on n and ℓ together, Eq. (4.5) can be rewritten and simplified as

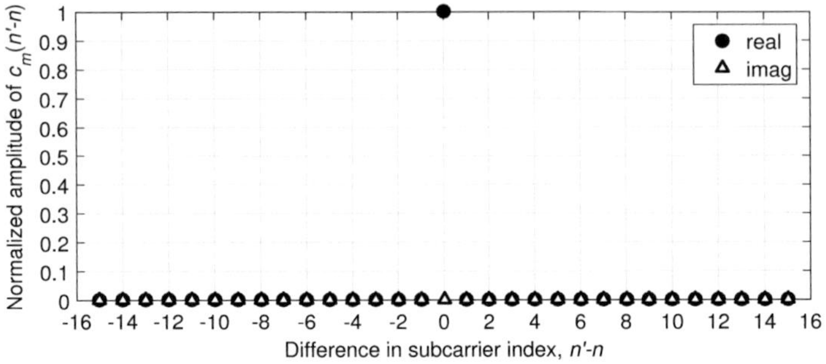

(a) Complex weights at $\delta_w T_0 = 0$. The amplitude of the symbol-under-evaluation $c_m(0) = 1$ does not leak to other subchannels.

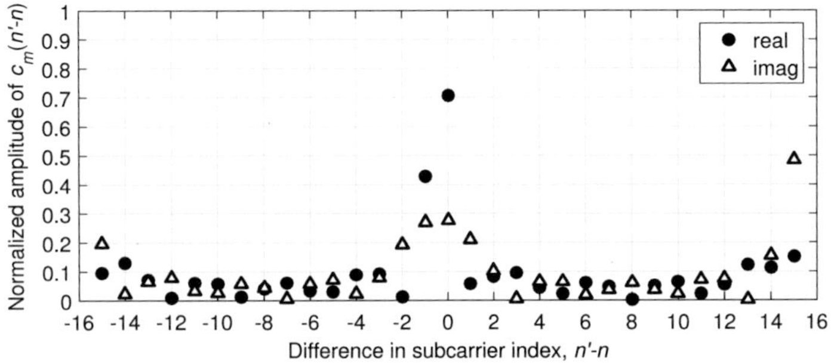

(b) Complex weights at $\delta_w T_0 = 0.4$. The amplitude of the symbol-under-evaluation $c_m(0) = 0.7$, with the rest leaking into other subchannels.

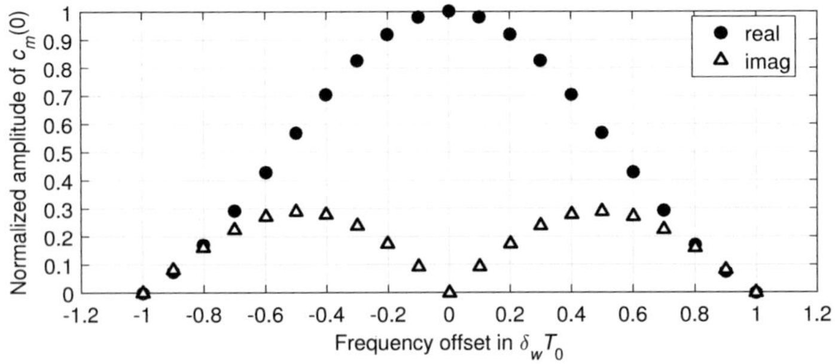

(c) Complex weight $c_m(0)$ at various amount of $\delta_w T_0$. The larger the frequency offset, the less the contribution of the symbol-under-evaluation to its own bin.

Fig. 4.3: Normalized complex weights showing the recovery of a modulation symbol $c_m(0)$ for various cases for an OFDM system with $N = 16$ subcarriers.

$$\begin{aligned}\tilde{a}_m(n') &= \mathrm{e}^{\mathrm{j}(\Psi+\zeta)}\,\mathrm{e}^{\mathrm{j}2\pi\delta_w mT}\sum_{n=0}^{N-1} a_m(n)\,\mathrm{e}^{-\mathrm{j}2\pi n\Delta f\tau}\sum_{\ell=0}^{N-1}\mathrm{e}^{\mathrm{j}\frac{2\pi}{N}\ell(n-n'+\delta_w T_0)} \\ &= \mathrm{e}^{\mathrm{j}(\Psi+\zeta)}\,\mathrm{e}^{\mathrm{j}2\pi\delta_w mT}\sum_{n=0}^{N-1} c_m(n-n')\cdot a_m(n)\,\mathrm{e}^{-\mathrm{j}2\pi n\Delta f\tau},\end{aligned} \tag{4.6}$$

where $c_m(n'-n)$ is the finite summation over a geometric exponential series of the last summation term over ℓ, and normalized by dividing over N so that their maximum amplitude is 1. This is called the 'complex weights' and is described by the following closed form expression

$$c_m(n-n') = \frac{1}{N}\frac{\sin\left(\pi\left(n-n'+\delta_w T_0\right)\right)}{\sin\left(\frac{\pi\left(n-n'+\delta_w T_0\right)}{N}\right)}\cdot\mathrm{e}^{\mathrm{j}\,\frac{N-1}{N}\left(n-n'+\delta_w T_0\right)}. \tag{4.7}$$

The complex weights describe the amount of energy leakage (in terms of a normalized factor of 0 to 1) from the symbol-under-observation $c_m(0)$ to its neighboring subchannels. From Eq. (4.6) and Eq. (4.7), it can be seen that if $\delta_w T_0 = 0$ when $n' = n$, the estimated symbol $\tilde{a}_m(n')$ will simply be the originally transmitted symbol with channel effects.

In order to visualize the effect of the frequency shift to the recovery of the modulation symbol, an $N = 16$ OFDM subcarrier system of the complex weights is plotted. When there is no frequency offset i.e. $\delta_w T_0 = 0$, the plot in Fig. 4.3(a) results. At $c_m(0)$ only one peak of real value can be seen, meaning that the DFT yields only one value for the symbol-under-evaluation. The neighboring bins (where $n' - n \neq 0$) all possess zero amplitudes, implying energy from $c_m(0)$ does not leak into its neighboring subchannels.

When a frequency shift of $\delta_w T_0 = 0.4$ is applied to the signal, the effect is as shown in Fig. 4.3(b). The DFT bins no longer line up with the peaks of the received symbols hence orthogonality is lost. The amplitude of $c_m(0)$ is only 0.7, with the rest of the energy spilled over to all the neighboring bins, in amounts that are also shown. When all the normalized complex values from $0,..,N-1$ are summed, they will result in a value of $1+\mathrm{j}0$.

Fig. 4.3(c) depicts how the amount of frequency shift would affect the contribution of $c_m(0)$ to the recovered modulation symbol. Typically, the larger the amount of frequency offset, the more the channel leakage to the other subchannels. At $\delta_w T_0 = \pm 1$, the orthogonality returns but $c_m(0) = 0$ because the symbol-under-observation is now fully contained in the adjacent subcarrier. Suffice to say, that the greater the frequency shift of the signal, the less the energy the symbol-under-evaluation has and the more it is phase-shifted from its original position in the constellation diagram.

4.1.4 Comm profile on radar signal

Taking into account all the channel and hardware effects that have been described for the Comm node, an expression for the Comm's signal at the Radar's receiver can now be written. For simplicity, assume here that the Comm has only one transmitter and one propagation path hence at the Radar, the Comm signal that is seen is

$$\begin{aligned}\mathbf{W}(n_u^w, m) =& \mathbf{S}_w(n_u^w, m)\,\alpha_w\,\mathrm{e}^{\mathrm{j}(\Psi_w+\zeta_w)}\,. \\ & \overbrace{\left[\mathrm{e}^{\mathrm{j}\varphi_{\mathrm{R}_w}(u^w)}\,\mathrm{e}^{-\mathrm{j}2\pi i N_{\mathrm{ch}}\Delta f \tau_w}\right]}^{\kappa_{r_w}(n_u^w)}\overbrace{\left[\mathrm{e}^{\mathrm{j}\varphi_{\mathrm{train},w}}\,\mathrm{e}^{\mathrm{j}2\pi\delta_w\left(mT+n_u^w T_{\mathrm{S}}\right)}\right]}^{\kappa_{\mathrm{D}_w}(n_u^w,m)},\end{aligned} \tag{4.8}$$

where $\mathbf{S}_w(n_u^w, m)$ are the original transmitted symbols of Comm and n_u^w is the subcarrier set assigned to the Comm. Any term designated with a subscript of an italic 'w' i.e. $[\cdot]_w$ refers to the Comm, whereby

- α_w is the amplitude attenuation of the signal due to the Euclidean distance r_w of the Comm node from the Radar. The amplitude attenuation is calculated using the Friis equation in Eq. (4.1).

$$\alpha_w = \frac{\lambda_\mathrm{c}}{4\pi r_w} \tag{4.9}$$

- $\mathrm{e}^{\mathrm{j}\Psi_w}$ is the phase offset experienced by the Comm signal due to r_w,

$$\Psi_w = -2\pi\frac{r_w}{\lambda_\mathrm{c}}. \tag{4.10}$$

- τ_w is the time delay of the signal due to it's propagation distance.

$$\tau_w = \frac{r_w}{c_0} \tag{4.11}$$

- δ_w is the summation of the Doppler and hardware CFO of Comm.

$$\delta_w = f_{\mathrm{LO},w} + f_{\mathrm{D},w} \tag{4.12}$$

- $\mathrm{e}^{\mathrm{j}\zeta_w}$ is the arbitrary random phase offset imposed by the channel and/or hardware.

Hence the received modulation symbols matrix at the q-th receive antenna of the Radar, along with the Comm's symbols modified from Eq. (2.38) can be written as

$$\begin{aligned}\mathbf{Y}_q(n,m) = \sum_{p=0}^{P-1}\mathbf{X}_p(n_u,m)\sum_{h=0}^{H-1}\alpha_h \mathrm{e}^{\mathrm{j}(\Psi_{p,q,h}+\zeta_{n_u,m})}\Big(\mathrm{e}^{\mathrm{j}\varphi_{\mathrm{R}_h}(u)}\,\mathrm{e}^{-\mathrm{j}2\pi(iN_{\mathrm{ch}})\Delta f\tau_h}\Big)\cdot \\ \Big(\mathrm{e}^{\mathrm{j}\varphi_{\mathrm{train}}}\,\mathrm{e}^{\mathrm{j}2\pi f_{\mathrm{D},h}(mT+n_uT_{\mathrm{S}})}\Big) + \mathbf{W}(n_u^w,m) + \hat{\mathbf{Z}}(n,m).\end{aligned} \tag{4.13}$$

Thus it can be seen here that the Radar's quotient matrix (see Eq. (2.41)) for the estimation of the range and Doppler and subsequently the azimuth and elevation, contains the Comm's symbols as well as noise. If the subcarrier set assigned to both the Radar and the Comm is different i.e. $n_u \neq n_u^w$ and $\delta_w = 0$ then the term $\mathbf{W}(n_u^w,m)$ will not interfere with the desired signals in Eq. (4.13). However, should there be any Doppler or CFO, $\mathbf{W}(n_u^w,m)$ will appear as white noise in the Radar's radar images that consequently lowers its $\mathrm{SNIR}_{\mathrm{out}}$.

4.2 OFDM frame for payload data recovery

Now that the Radar's and Comm's signal characteristics have been defined, this section will proceed with the discussion about the additional headers or symbols that must be added to the RadCom's signal for payload data recovery. As has been discussed in the previous section, any frequency offset at the Comm's signal will rotate its modulation symbols away from their correct locations in the constellation diagram. Since any time delay of the signal also incurs a phase rotation, this will also add to the overall phase rotation. The demodulation of these symbols are very dependent on the area of the constellation diagram, which they lay. For instance, for a 4-PSK type modulation, the constellation diagram will be divided into 4 equal quadrants. Each quadrant represents a modulation alphabet and each symbol that is within that particular quadrant will be demodulated accordingly. Thus when a modulation symbol has been rotated by a sufficient amount into another quadrant, its demodulation will also be erroneous.

In order to recover the modulation symbols correctly, they must first be shifted back to their original positions by the correct amount. Since the phase rotations are caused by the time delay and frequency offsets, a time and frequency synchronization must be performed on the received signal before they are put through the demodulation process. OFDM synchronization techniques are divided into - 1) non data-aided, and 2) data-aided. The former has better efficiency or throughput due to having no redundant data but the performance is not sufficiently robust in a multipath environment. An example of this technique uses the CP for time and frequency estimation and is presented in [vdBSB97], but the bad performance is confirmed in [CMB08]. The latter data-aided technique requires some redundancy in the form of preambles or training symbols and pilot tones and has better synchronization capabilities and robustness, which come at the expense of less efficient data throughput. One of the best and most popular methods is the Schmidl and Cox (SCA) [SC97] method, which has sprouted many spin-off versions such as Morelli's [MM99] and Minn's [MZB00] algorithms.

Since the optimization of time and frequency offset synchronization is not the main goal of this thesis, the SCA will be used, along with pilot tones. These will be discussed in the following sections.

4.2.1 Schmidl and Cox algorithm

The SCA is specifically designed for OFDM time and frequency synchronization for use in communication applications. It is very popular due to its accuracy, which is close to the Cramer Rao bound. The SCA uses two training symbols for synchronization. The first symbol, denoted as $c_{1,n}$, is used to estimate the symbol start time and fractional frequency offset that is within $[-\frac{\Delta f}{2}, \frac{\Delta f}{2}]$. The second symbol, denoted as $c_{2,n}$, is used to gauge the integer frequency offset for when the frequency offset exceeds one subcarrier spacing. To avoid confusion, the term '**training symbol**' is used when referring to the frequency domain SCA symbols and '**preamble**' for the time domain equivalent.

Fractional frequency offset estimation

The first training symbol $c_{1,n}$ is generated with pseudo-noise (PN) sequences using a random selection of modulation symbols within $\mathcal{A}$ only on the even-index subcarriers while the odd-index subcarriers are not used and are set to zero. An example is shown in Table 4.1. This is in essence a zero-interpolation between samples in the frequency domain hence the time domain equivalent will repeat at twice the rate yielding two identical halves. Just like the OFDM symbol, a CP is also added to the preambles. To estimate the fractional frequency offset, a comparison of two identical halves of the preamble is made. These halves separated by $\frac{N}{2}$ samples incur the phase rotation as given by the first term in Eq. (4.4) and will have approximately the same timing offset. As such, the phase difference between the symbols in the first and second half is

$$\begin{aligned} \phi_{\text{frac}} &= \phi_{d+\frac{N}{2}} - \phi_d = 2\pi\left(n+\frac{N}{2}\right)\delta\frac{T_0}{N} - 2\pi n\delta\frac{T_0}{N} = \pi\delta T_0 \\ \Rightarrow \delta &= \frac{\phi_{\text{frac}}}{\pi T_0}, \end{aligned} \tag{4.14}$$

where δ is the frequency offset that shifts the spectrum by a constant amount within $\phi_{\text{frac}} = [-\pi, \pi]$ hence it is called the 'fractional' frequency offset. This is equivalent to a frequency offset of an amount that is within the subcarrier spacing (analogous to Eq. (2.20)) where the maximum frequency offset is $\delta_{\max} = \pm\frac{1}{T_0}$. The term ϕ_{frac} can be estimated by doing an XCORR with a sliding window of size d to obtain the timing metric $P(d)$ (similar to an XCORR output) as given by

$$\begin{aligned} P(d) &= \sum_{\ell=0}^{\frac{N}{2}-1} r^*_{d+\ell}\, r_{d+\ell+\frac{N}{2}} \qquad \ell = 1, \ldots, \frac{N}{2} \\ P(d+1) &= P(d) + \left(r^*_{d+N/2}\, r_{d+N}\right) - \left(r^*_d\, r_{d+N}\right), \end{aligned} \tag{4.15}$$

where r_ℓ and $r_{\ell+\frac{N}{2}}$ are the first and second half of the preamble. The timing metric is found by normalizing $P(d)$ by $R(d)$ to yield the normalized timing metric $M(d)$,

$$M(d) = \frac{|P(d)|^2}{R(d)^2} \quad \text{with} \quad R(d) = \sum_{\ell=0}^{\frac{N}{2}-1} \left|r_{d+\ell+N/2}\right|^2, \tag{4.16}$$

where $R(d)$ is the received energy for the second half of the time symbol and $M(d)$ is a multiple-peak XCORR plot with the peaks forming a quasi-plateau. Ideally, the peaks over the d sliding window should be of the same amplitude but this is rarely the case when there are multipath signals present. Therefore the timing offset is taken as the lag or time sample index d^* corresponding to the maximum point of $M(d)$, which is a point on the plateau as the most probable signal start point. This also means that the estimated start point can fall anywhere within the CP length of the signal or more. The estimated phase difference is then the angle of the highest peak $\angle P(d^*)$, and the frequency offset $\hat{\delta}$ is simply given by

$$\hat{\phi}_{\text{frac}} = \angle P(d^*) \qquad \Rightarrow \hat{\delta} = \frac{\angle P(d^*)}{\pi T_0}. \tag{4.17}$$

Table 4.1: Example SCA training symbols.

subcarrier index, n	$c_{1,n}$	$c_{2,n}$	$v_n = \sqrt{2}\frac{c_{2,n}}{c_{1,n}}$
0	1.0000 − j1.0000	−0.7071 + j0.7071	−1
1	0	−0.7071 − j0.7071	
2	−1.0000 − j1.0000	−0.7071 − j0.7071	1
3	0	0.7071 − j0.7071	
4	−1.0000 + j1.0000	−0.7071 − j0.7071	j
5	0	−0.7071 + j0.7071	
6	−1.0000 − j1.0000	−0.7071 + j0.7071	−j
7	0	0.7071 − j0.7071	
8	1.0000 + j1.0000	0.7071 + j0.7071	1
9	0	0.7071 + j0.7071	

Integer frequency offset estimation

When the ϕ_{frac} is larger than $[-\pi, \pi]$, this second part of the algorithm then estimates the frequency offset equivalent to the number of subcarrier shifts. This is found by comparing both training symbols (c_1 and c_2) using only the PN-sequence on the even-index subcarriers. Let x_1 and x_2 be the DFT of the first and second received preambles. Defining $x_{1,k}$ and $x_{2,k}$ as the symbols occupying the even-index subcarriers, where $k = 0, 2, 4, ..., N-2$, the phase shift between $x_{1,k}$ and $x_{2,k}$ is

$$\phi_{\text{int}} = \phi_{x_{2,k}} - \phi_{x_{1,k}} = 2\pi(g2\Delta f)\,2T - 2\pi(g2\Delta f)\,T = 2\pi\frac{2g}{T_0}T, \tag{4.18}$$

where $2\Delta f$ is due to the even-index only subcarriers used and g is the integer amount of shift. To find $\frac{2g}{T_0}$ the fractional frequency offset must first be partially corrected to avoid ICI. This is done by multiplying the received time domain signal by the inverse of the estimated phase offset $e^{-j2\pi\frac{\phi_{\text{frac}}}{T_0}t}$. Similar to the timing metric, g can be estimated by finding the $\hat{g}$ that maximizes the XCORR between the first and second received training symbols with

$$B(g) = \frac{\left|\sum_{k} x^{*}_{1,k+2g}\, x_{2,k+2g}\, v^{*}_{k}\right|^{2}}{2\left(\sum_{k} |x_{2,k}|^{2}\right)^{2}}, \tag{4.19}$$

where the denominator is the energy of the second received training symbol and v_k is the phase correction factor of the $c_{2,k}$ and $c_{1,k}$ symbols. The operation in the numerator will equalize all phase differences due to the different symbols and leave only the amplitude and ϕ_{int}. Sliding through the window over the variable g, the index that sees the highest value at the numerator is the most likely estimator of $\hat{g}$. The integer frequency offset is then estimated as $\frac{2\hat{g}}{T_0}$. Hence the total frequency offset estimated will be

$$\hat{\delta}_{\text{total}} = \frac{\hat{\phi}_{\text{frac}}}{\pi T_0} + \frac{2\hat{g}}{T_0}. \tag{4.20}$$

4.2.2 Pilot tones

In addition to the SCA preambles, which are prepended at the beginning of the time domain transmit signal, pilot tones are also embedded within the transmit frame to track the changes in the channel. This will then allow the equalizer at the Rx to construct the channel equalization matrices to reverse those effects. The pilot tones, like the training symbols, are arbitrary symbols but known *a priori* to all communication partners, and is a form of redundancy. There are two basic types of 1D channel estimation based on the pilot tone distribution as shown in Fig. 4.4 as explained by Shen and Martinez in [fre20]. The block type in Fig. 4.4(a) is sufficient to counter slow fading while the comb type in Fig. 4.4(b) is to counter frequency-selectivity in the channel.

To take both the change in the time and frequency domains, a hybrid of the block and comb type pilot tones is used. In the simplest form, these tones are spread regularly within the OFDM transmit frame in accordance to the Nyquist sampling theorem as shown in Fig. 4.5(a). Although there are improved channel estimation based on pilot tone distribution and specific estimator algorithms for wireless OFDM communication systems such as presented in [OA07, Cav91], they are not within the scope of this work. Hence throughout this thesis, steps are taken to ensure that:

- the delay spread of the signal is smaller than or equal to the OFDM symbol duration. This ensures that the signal bandwidth is always within the coherence bandwidth and the signal only undergoes flat fading. To avoid inter-symbol interference (ISI) for the Radar, the maximum round-trip time from the farthest target is $\tau_{\mathrm{max}} \leq T_0$.
- the Doppler spread of the signal is always within the subcarrier spacing, which confines the maximum frequency offset is $f_{\mathrm{D_{max}}} < \Delta f$. This ensures slow fading.

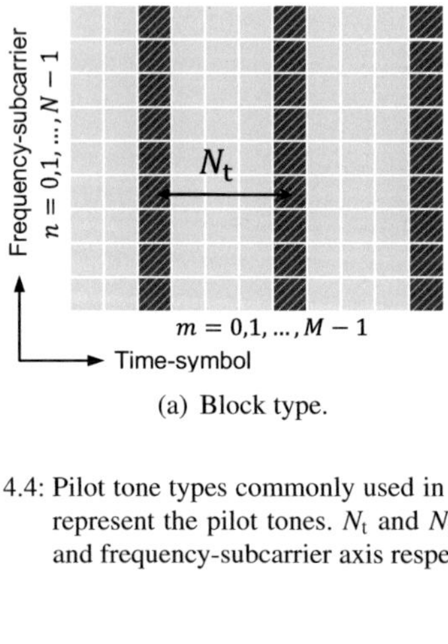

(a) Block type.

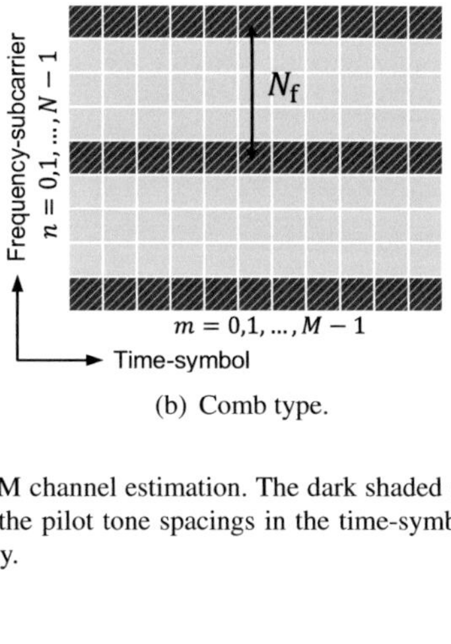

(b) Comb type.

Fig. 4.4: Pilot tone types commonly used in OFDM channel estimation. The dark shaded squares represent the pilot tones. N_{t} and N_{f} are the pilot tone spacings in the time-symbol axis and frequency-subcarrier axis respectively.

To determine the pilot tone spacing in the time-symbol axis, the maximum unambiguous Doppler term in Eq. (2.18) will be utilized. Incorporating the sampling theorem, the maximum expected frequency shift of the scenario, $f_{\mathrm{D_{max}}}$ can be written as

$$f_{\mathrm{D_{max}}} = \frac{1}{2T \cdot N_{\mathrm{t}}} \quad \Rightarrow \quad N_{\mathrm{t}} \leq \frac{1}{2 f_{\mathrm{D_{max}}} T}, \tag{4.21}$$

where $\frac{1}{N_t}$ can be seen as the sampling rate or N_t being the spacing (in terms of number of bins) between two pilot tones in the time-symbol axis of the OFDM frame. Analogously using Eq. (2.16), the maximum expected time delay or range affecting the frequency-subcarrier axis can be written as

$$\frac{1}{\Delta f N_{ch}} \frac{c_0}{2} = \tau_{max} \cdot 2N_f \frac{c_0}{2} = R_{max} \quad \Rightarrow \quad N_f \leq \frac{1}{2N_{ch}\Delta f \tau_{max}}, \tag{4.22}$$

where R_{max} is the maximum expected range with the time delay τ_{max} of the Comm in the scenario and N_f is the minimum spacing between bins in the frequency-subcarrier axis. While a smaller pilot tone spacing can resolve larger time delay and frequency offset values, they compromise the available leftover bins in the transmit frame meant for the payload data. Thus the pilot tone distribution should be chosen to be just adequate to adapt to the maximum foreseeable time delay and frequency offset in the scenario to minimize the data redundancy.

4.2.3 OFDM transmit frame

Putting together the training symbols and pilot tones into the OFDM transmit frame, the size of the frame will increase to $(N + N_{train}) \times M$ as shown in Fig. 4.5(a). Note that the pilot tones will also line the last row and column of the OFDM frame irrespective of the indices given by N_f and N_t. This surrounds the OFDM frame with pilot tones to enable better interpolation (to be discussed in Chapter 5) to obtain the channel equalization matrix. The squares that are blank can be filled with any arbitrary data such as from a text or graphic, which have been coded with a modulation alphabet.

The training symbols are as shown in Fig. 4.5(b). They are generated with one missing symbol at the upper right side of the constellation diagram for easier identification by sight as will be apparent in the next chapter. This can easily be done by generating the PN sequence using only three out of four possible constellation points. The points representing the symbols have also been made bigger for better visibility. To make the payload data more robust to the channel influences, forward error correction codes i.e. source and channel coding can also be used with no consequence to the radar's estimation algorithm. In this thesis, no coding is used to eliminate their influence on the Radar's performance in terms of bit error ratio (BER) as will be discussed in Chapter 5.

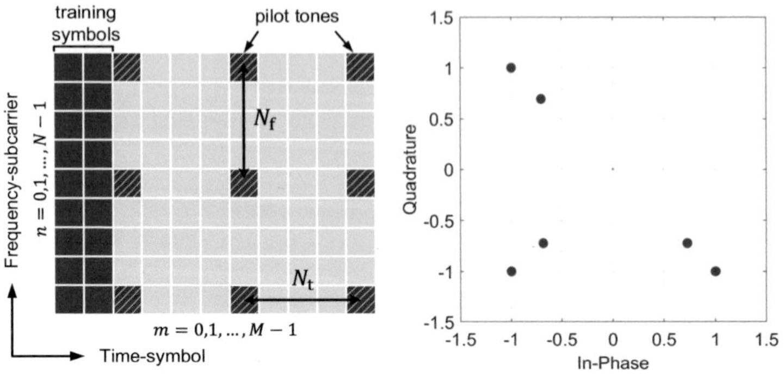

(a) OFDM frame.

(b) SCA training symbols.

Fig. 4.5: (a) OFDM frame with prepended preamble and regularly spaced pilot tones according to N_f and N_t. The pilot tones also surround the borders of the frame. (b) SCA training symbols. The top right side of the constellation is generated void of training symbols and the points representing the symbols have been made bigger, for better visibility.

4.3 Mutual interference tolerance of the RadCom

Now that the characteristics of the Comm node is known along with its effect on the Radar's receive signal, the next step is to obtain the quantitative effect of interference on the Radar's $\mathrm{SNIR_{out}}$. The work presented in [SSBZ13] shows that the channel isolation (in terms of $\mathrm{SNIR_{out}}$) remains above 70 dB for a frequency offset of up till $\delta_w T_0 = 0.0495$. This isolation value is however only valid without any noise (i.e. $\mathrm{SNR} = \infty$ in simulation) or taking only the noise floor of the measurement instruments into account ($\mathrm{SNR} > 30\,\mathrm{dB}$), together with specific considerations of the scenario and hardware's range of operations. Essentially, for a radar only scenario without any interferer, the $\mathrm{SNR_{out}}$ is dependent on the amount of power backscattered from the targets and the amount of Doppler incurred. Logically this means that the farther a target is from the radar, the more path loss the backscattered signal will undergo. Moreover the RCS also determines the amount of power that is backscattered. When there is a Doppler shift, the less than maximum contribution from the desired symbol has the same effect as raising the noise floor. Hence the maximum achievable $\mathrm{SNR_{out}}$ is a variable that is dependent on the scenario.

4.3.1 Simulation scenario setup

Here, it is of interest to see how the power as well as the frequency shift of the Comm affects the Radar's $\mathrm{SNIR_{out}}$, a relation, which has not been previously explored. A scenario with six targets according to Table 4.2 is set up in Matlab. The 4-PSK modulation is used. The other OFDM parameters are as shown in Table 5.2, but the outcome of this simulation is not influenced by the aforementioned parameters. The signal bandwidth is divided into $N_{\mathrm{ch}} = 8$ channels, with the Radar occupying the subcarrier set given by $u = 0$ and the Comm, $u = 1$ as shown in Fig. 4.6. This was chosen based on the worst-case scenario subchannel occupation results in [SSBZ13]. The other channels are unoccupied. The Comm power in terms of $\mathrm{SIR_{in}}$ is varied along with the frequency offset $\delta_w T_0$. A Hamming window is used for all radar images[15], thereby reducing the maximum $\mathrm{SNIR_{out}}$ by approximately 2.7 dB. Based on the setup of $N/N_{\mathrm{ch}} = 128$ and $M = 256$, the $G_{\mathrm{per}} = 45.15$ dB, and through simulation, the maximum achievable $\mathrm{SNIR_{out}}$ in this scenario is determined to be 77 dB (with an SNR of ∞ dB or no added white noise).

Table 4.2: Target parameters for mutual interference tolerance simulations.

Target index	Distance in m	Relative velocity in m/s	RCS in m^2	Azimuth in °
A	20	0	10	5
B	35	−20	10	25
C	40	10	10	-15
D	50	25	10	-30
E	70	−25	10	-10
F	90	−25	10	35

The chart in Fig. 4.7 shows the $\mathrm{SNIR_{out}}$ (given in terms of normalized intensity in dB) over varying $\mathrm{SIR_{in}}$ of −50 to 0 dB and frequency offset of $-1 \leq \delta_w T_0 \leq 1$. The first observation that can be made is that as the Comm's frequency offset becomes more negative (going toward the direction of the Radar's subchannel), the $\mathrm{SNIR_{out}}$ decreases by 35 dB even when the $\mathrm{SIR_{in}}$ remains at 0 dB. This is in line with theoretical expectations. As the $\mathrm{SIR_{in}}$ decreases (becomes more negative), the $\mathrm{SNIR_{out}}$ is affected even more severely. In fact, the $\mathrm{SIR_{in}}$ affects the $\mathrm{SNIR_{out}}$ more rapidly than the increase of the frequency offset. When the frequency offset moves toward the positive values (toward an unused subchan-

[15] for Hamming windowing over the frequency-subcarrier axis and the time-symbol axis, hence $2 \cdot 1.36\,\mathrm{dB} = 2.7$ dB, refer to Chapter 2.5.5.

nel), the $\mathrm{SNIR_{out}}$ decreases at first to the point where $\delta_w T_0 = 0.5$ and then increases again as the amount of subcarrier misalignment also decrease.

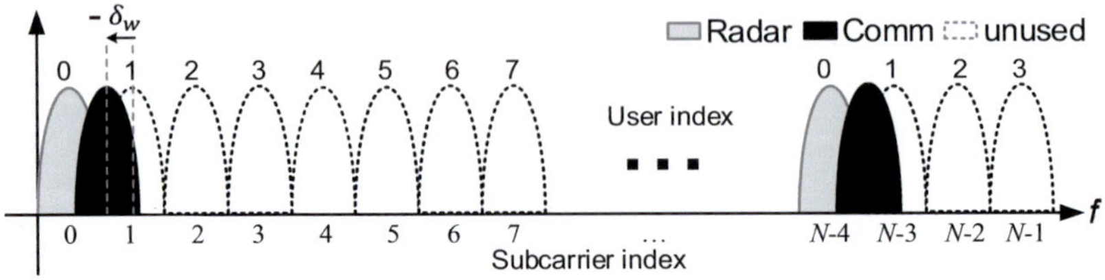

Fig. 4.6: Subcarrier assignment for the Radar $u = 0$ and Comm $u = 1$ for simulation to result in Fig. 4.7.

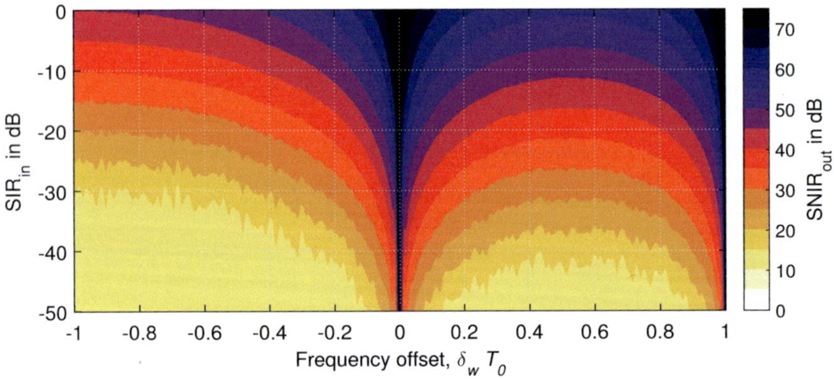

Fig. 4.7: Simulated radar tolerance chart based on the worst-case scenario of direct neighbouring channel assignment in [SSBZ13] for Comm (interferer) according to Fig. 4.6. The maximum achievable $\mathrm{SNIR_{out}}$ with 45.15 dB processing gain and 2.7 dB Hamming window loss is around 77 dB. Each color level represents a 5 dB step.

It can be seen that the $\mathrm{SNIR_{out}}$ value plateaus at 10 dB in Fig. 4.7. By the definition of the $\mathrm{SNIR_{out}}$, the strongest target peak is normalized to be 0 dB and hence the absolute value of the normalized average noise floor becomes the $\mathrm{SNIR_{out}}$ value. The noise peak however can range from 10 to 12 dB above the average noise floor hence when the $\mathrm{SNIR_{out}}$ of the radar image is around 10 dB, it can be interpreted as the strongest target peak having the same or lower amplitude than the noise peak.

In order to reliably distinguish the target peak from the noise peak, the target peak should be substantially higher than the highest noise peak, which puts the required $\mathrm{SNIR}_{\mathrm{out}}$ at approximately 15 to 20 dB. A constant false alarm rate (CFAR) algorithm (first described by Finn in [Fin86] and elaborated by Bunch and Piero in [BF92]), which sets a threshold to the radar's received signal to distinguish noise and clutter from the desired targets can be implemented to improve the reliability of estimation. The CFAR has spawned many improved versions for target detection under severe interference such as presented in [Cao04]. This will however not within the scope of this work and will not be further elaborated.

To utilize this chart for another scenario, the maximum $\mathrm{SNIR}_{\mathrm{out}}$ in the chart must first be scaled to the maximum achievable $\mathrm{SNIR}_{\mathrm{out}}$ in the new scenario, which is also dependent on the transmit power, target distance and SNR of the Radar. This chart can be used in two cases. Case 1 is when a Comm with known power and frequency offset is applied either in simulation or measurement. By looking up its $\mathrm{SIR}_{\mathrm{in}}$ and $\delta_w T_0$ on the y- and x-axis of the chart, the expected achievable $\mathrm{SNIR}_{\mathrm{out}}$ range can be estimated. This expected $\mathrm{SNIR}_{\mathrm{out}}$ is to be applied to the furthest target in the scenario. Case 2 is when a Comm with unknown power or frequency offset is in the vicinity, a scenario that is mostly encountered during real measurements. Based on the Radar's $\mathrm{SNIR}_{\mathrm{out}}$ outcome, the unknown $\mathrm{SIR}_{\mathrm{in}}$ and $\delta_w T_0$ Comm parameters can also be narrowed down to a few possibilities. This information can then be used to deduce whether the radar estimation should proceed or must first be put through an interference cancellation algorithm.

To demonstrate the severity of the Comm signal interfering on the Radar, the Radar is allocated $u = \{0, 1, 2, 3\}$ and the Comm $u = 4$. The transmit power of the Radar and the Comm are the same and the $\mathrm{SIR}_{\mathrm{in}}$ is set to -30 dB (equivalent to the Comm being 12.7 m away from the Radar) with $\delta_w T_0 = 0.16$. To make the scenario more realistic, an SNR of 30 dB at the Radar's Rx is used. The resulting radar image is taken at $u = 3$ of Rx 1 is as shown in Fig. 4.8(a). Due to the normalized noise floor (inclusive of interference from Comm) of around -26 dB only the first three targets nearest to the radar (Target A, B and C from Table 4.2) can be identified, where Target A, B and C are at 26 dB 15 dB and 12 dB respectively above the noise floor. Without using additional post-processing algorithms such as the CFAR, the estimation of Targets B and C are unreliable.

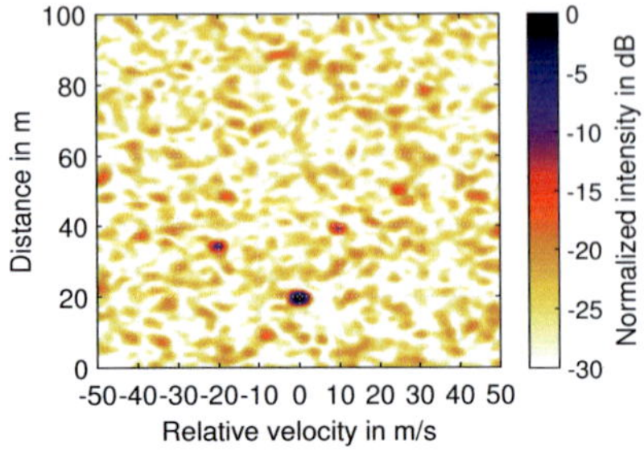

(a) Radar image from $u = 3$ of Rx 1. $\text{SNIR}_{\text{out}} = 26\,\text{dB}$.

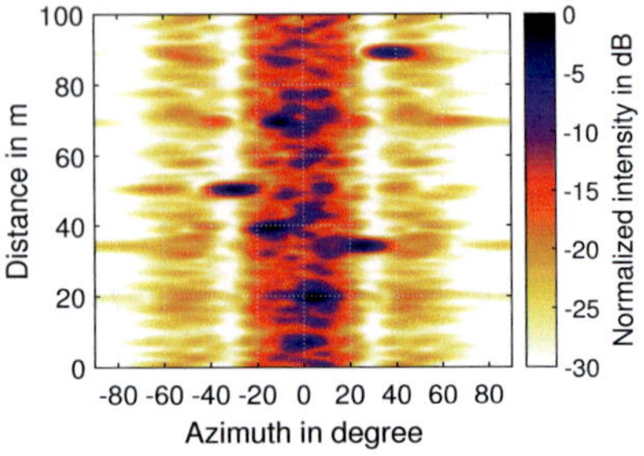

(b) DOA image (summed).

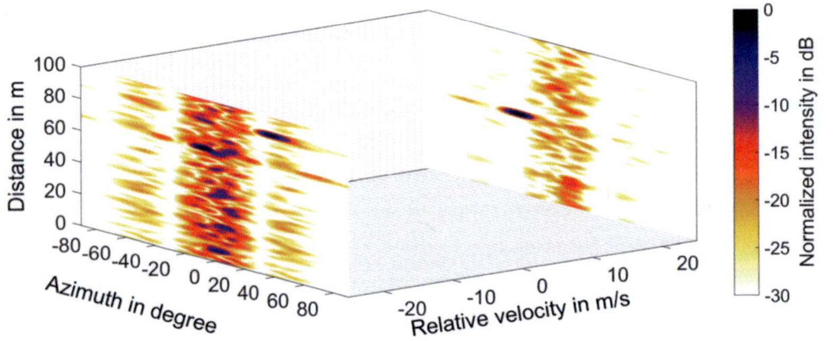

(c) DOA image (slice) of Targets E and F at −25 m/s and D at 25 m/s.

Fig. 4.8: Simulated radar image with one Comm signal at 12.70 m away from Radar ($\text{SIR}_{in} = -30\,\text{dB}$), $\delta_w T_0 = 0.16$ and SNR = 30 dB. Only the first three targets (A to C) out of six can be seen in the radar image. The SNIR_{out} is 26 dB at $u = 3$.

The corresponding DOA matrices are processed according to the method described in Section 3.4. This results in a 3D matrix of different range-azimuth 'DOA slices' according to their velocities. The superimposed or 'summed' DOA image of all targets are as shown in Fig. 4.8(b), processed using a simple summation of the five DOA image slices containing the six targets. This method, although simple for detecting all unknown targets in the scenario also increases the noise floor by around 14 dB (for this particular scenario only) hence the SNR of the image will be lower than expected.

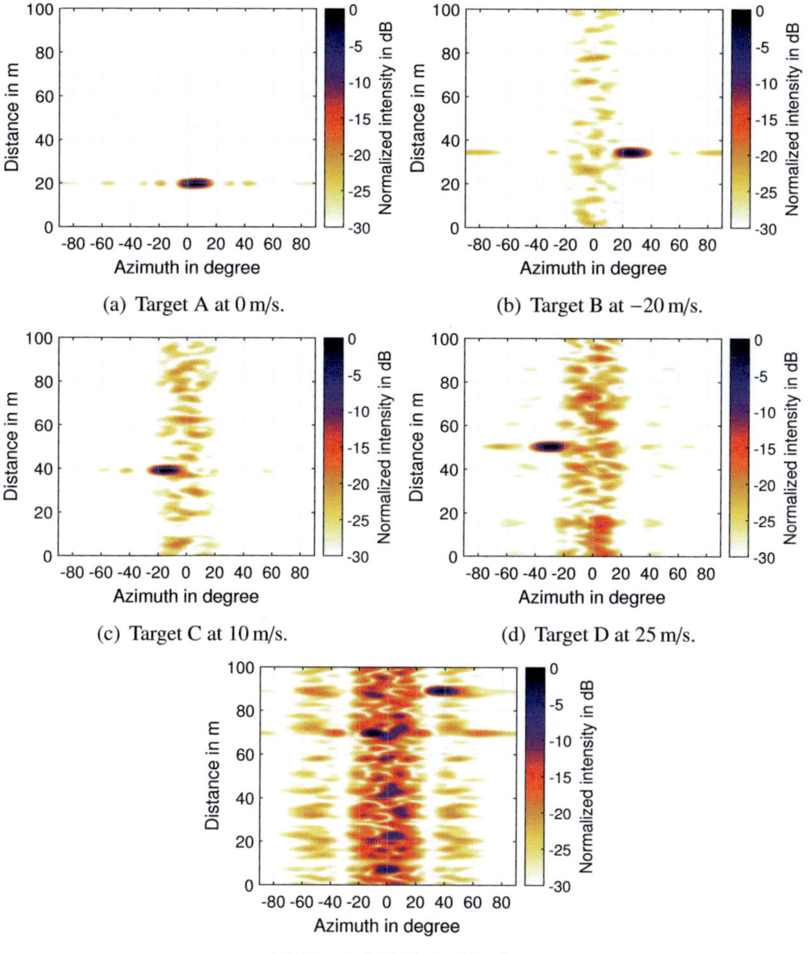

(a) Target A at 0 m/s.

(b) Target B at −20 m/s.

(c) Target C at 10 m/s.

(d) Target D at 25 m/s.

(e) Targets E & F at −25 m/s.

Fig. 4.9: Simulated DOA image slices based on velocity according to the scenario given in Fig. 4.8.

The DOA image slices for Targets E and F (at −25 m/s) and Target D (at 25 m/s) are demonstrated in Fig. 4.8(c). In these images, the target peaks are about $20\log_{10}(P \cdot Q) = 24\,\mathrm{dB}$ higher than all their corresponding target peaks of the radar image, due to the constructive summation of coherent phases during the Fourier Beamforming processing. The noise floor in all DOA image slices however is uncorrelated hence its increase can be approximated by $10\log_{10}(P \cdot Q) = 12\,\text{dB}$. For an ideal environment where there is no clutter, the target peaks of the DOA images will have a net gain of $10\log_{10}(P \cdot Q) = 12\,\text{dB}$. In the measurements however, clutter will also reflect and contribute to the increase of the noise floor in the DOA matrix. This is especially obvious at zero Doppler since typical clutter tend to be static, as will be demonstrated in Section 7.5.1. As a result the net gain may fall below 0 dB. All DOA images have sidelobes appearing as the horizontal trails along the targets arising from the sidelobes of the virtual radiation pattern, which were not adequately suppressed by the very short Hamming window of length $P \cdot Q$.

4.4 Chapter 4 summary

A loss of perfect OFDM subcarrier orthogonality is unavoidable in real scenario, where there are hardware imperfections (creating non-linear distortions) and moving RadCom nodes and targets. This loss of orthogonality is caused by a frequency offset that leaks the energy of symbols from all other subcarriers in the signal bandwidth into the desired symbol. When this happens, the Radar sees this leakage as an increased level of white noise. The higher the power of the Comm (interferer) signal and the frequency offset, the higher this perceived level of noise. Consequently the SNIR_{out} of the Radar also suffers.

Since the Comm signal power is almost always higher than the desired Radar's backscattered signal power, the tolerance chart in Fig. 4.7 can help determine if the achievable SNIR_{out} in a particular scenario will result in a reliable estimate. The next chapter will explore an interference cancellation method that addresses the issue of the unreliable estimates that have been demonstrated in this chapter.

5 Interference cancellation algorithm and analysis

Based on the observations of the radar tolerance chart in the previous chapter, it can be seen that the 'reliable estimation region', where the normalized $\mathrm{SNIR}_{\mathrm{out}} \geq 20\,\mathrm{dB}$, is limited to small $\mathrm{SIR}_{\mathrm{in}}$ values and small Comm frequency offsets $\delta_w T_0$. The $\mathrm{SIR}_{\mathrm{in}}$ value is very scenario dependent (on the range and RCS of the target and distance of the Comm). Therefore within the operations of networked radars the issue of estimation reliability becomes the ultimate challenge. This is especially true for OFDM-based systems since mutual interference cannot be removed via conventional adaptive filtering, just like how white noise cannot be filtered.

The research in [BTJ13] introduced a new metric for OFDM radar networks called the 'radar network outage probability' as a performance indicator. This metric describes the circumstances (in terms of probability, node per area size) when the radar network is too overwhelmed by interference to perform reliable target detection. It has been calculated that an OFDM system based on Sturm's and Wiesbeck's work in [SW11] has the probability of less than 1% for densities below one RadCom node per $158.5\,\mathrm{m}^2$. In a regular automotive or short-range area surveillance scenario, the radar node density per $158.5\,\mathrm{m}^2$ is foreseen to exceed at least two nodes, raising also the probability of a radar network outage substantially. Hence this becomes a strong motivation to seek a practical solution to mutual or inter-system interference.

Interference cancellation for OFDM-based radar is unlike the one for OFDM communication systems. The idea of interference cancellation in communication systems is to remove enough interference to recover the payload data. For the radar however, its goal is to separate the entire received interferer (Comm) signal from the desired useful signal that is reflected from the targets. The better the separation and subsequent cancellation of the interfering signal, the higher the $\mathrm{SNIR}_{\mathrm{out}}$ of the radar measurement. That being said, the process of separating the desired and undesired signals at the Radar's Rx is not straightforward. While regular OFDM demodulation techniques used in communication system can be applied to the Radar as well, these techniques have insufficient accuracy to correctly reproduce the interfering signal to be subtracted from the overall received signal. Hence additional steps must be taken to narrow the parameter deviation and these steps will be discussed in the following sections.

The idea behind this received signal separation of the desired and undesired signals is to first recover the payload data while estimating channel variables especially the time delay and frequency offset incurred to a considerably accurate level. The concept was first presented for a SISO OFDM RadCom configuration in [SSZ11,SSZ12] and then expanded for a MIMO configuration in [SZ14]. The correct recovery of the payload data directly affects the quality of the outcome. The recovered payload data is then remodulated with the estimated channel effects and subtracted from the received signal. Should there be more than one Comm present, this technique can be used in an iterative loop until a sufficiently high radar $\mathrm{SNIR}_{\mathrm{out}}$ is achieved.

5.1 Time synchronization

The first step to detecting any signal is to detect the start point of the signal. This is aided by the use of the preambles, which have been designed to give good auto-correlation peaks. The correct start point then leads to the correct sampling interval T_{S} to be applied for the retrieval of the signal. There are two main methods to detect the start point of the received OFDM symbols using preambles. The first method uses a simple XCORR. A list of peaks along with their time lags (or sample shifts) that are above a set threshold (determined through empirical or adaptive methods according to the scenario) will emerge. This list of peaks corresponds to the number of potential payload data signals. The XCORR will yield no other information other than the location of the peaks and their amplitude, posing a difficulty in selecting the time lag or index to use.

The second method uses the maximum likelihood approach such as used in the SCA presented in Section 4.2.1. As has been discussed in the previous chapter, the SCA also uses an XCORR method, albeit with a sliding window. The most likely start point is the d^* (see Section 4.2.1), which is the maximum point of the resulting quasi-plateau. The time lag of d^* can fall anywhere between the real start point of the symbol to the length of the CP and in the presence of multiple overlaid signals, might fall far beyond the signal length of $M \cdot T$. This incorrect removal of the Comm signal length will affect the its correct reconstruction and ultimately cause a degradation of the interference cancellation quality. As such, this method is not suitable to be implemented for radar interference cancellation. This is only the first block of payload data detection and the margin of tolerance between the recovery of a payload data for communication versus radar systems has already become apparent.

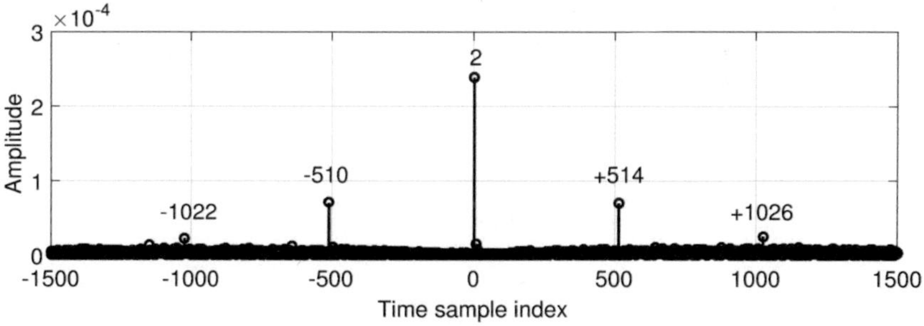

Fig. 5.1: Result of a double-sided cross-correlation between original preambles and received signal containing one payload data with $N = 1024$. Looking only at the positive time index axis, there are three consecutive peaks due to the SCA preambles (two identical halves in time domain) and cyclic prefix. The plot shows that the payload data signal has a time lag of 2 bins. Hence the start of the signal is at the time index of 3.

A modified XCORR function is presented here instead. By taking into account the preamble's characteristics, a better selection of the time lag can be made. The first SCA preamble consists of two identical halves along with a CP. When an XCORR between the original preambles and the received signal is made, three consecutive peaks spaced $\frac{N}{2}$ apart will emerge as shown in an example in Fig. 5.1, simulated with $N = 1024$. It can be seen that the signal containing the payload data has a time lag of 2 bins and the consecutive peaks are spaced 512 lags apart. To speed up the peak detection, a threshold can be used to suppress any peaks that are below the threshold value. This threshold value can be set adaptively based on the scenario. When there is more than one payload data signal in the received signal, more peaks will result. This modified XCORR function then searches for all peak pairs located $\frac{N}{2}$ apart. If a pair is found, the stronger peak with the smaller time lag is confirmed as originating from the preambles used and not from a random signal spike. All the detected peaks are then ranked by their order of strength (instead of time lag) and the interferer signals will be recovered and canceled accordingly by the listed order.

To correct the phase rotations due to distance it is easier to start sampling the time domain receive signal based on the estimated time lag rather than to compensate for the phase rotations in the frequency domain. Assuming a correct sampling interval, τ_w (from Eq. (4.8)) can only be estimated to the nearest integer given by

$$\lfloor \tau_w \rfloor = T_\mathrm{S} \cdot N_{\tau_w} \qquad N_{\tau_w} \in \mathbb{Z}. \tag{5.1}$$

The output of the XCORR is N_{τ_w}, based on the strongest peak with a consecutive pair. Even if the estimated time lag is correct to the nearest time bin, since N_{τ_w} is an integer, there will still be some residue phase rotations uncorrected. Let $\tilde{\tau}_w$ be the estimated time lag from the XCORR operation. The residue time delay can be represented as $(\tau_w - \tilde{\tau}_w)$. The time synchronization is done by shifting the start point of the time domain signal samples to $N_{\tau_w} + 1$. If $(\tau_w - \tilde{\tau}_w) < T_\mathrm{S}$, where the corresponding range residue is less than one range resolution, the time synchronization is considered accurate and the residue phase rotations will be corrected by the channel coefficients (presented in the next sections). The time domain received signal can now be written as in Eq. (4.13), with the time synchronized strongest modulation symbols (of the Comm) $\mathbf{W}(n_u^w, m)$ being

$$\begin{aligned} \widetilde{\mathbf{W}}^{\tau}(n_u^w, m) = \mathbf{S}_w(n_u^w, m)\,\alpha_w\,\mathrm{e}^{\mathrm{j}(\Psi_w + \zeta_w)} &\left[\mathrm{e}^{\mathrm{j}\varphi_{\mathrm{R}_w}(u^w)}\,\mathrm{e}^{-\mathrm{j}2\pi i N_\mathrm{ch}\Delta f(\tau_w - \tilde{\tau}_w)}\right]\cdot \\ &\left[\mathrm{e}^{\mathrm{j}\varphi_{\mathrm{train},w}}\,\mathrm{e}^{\mathrm{j}2\pi\delta_w\left(mT + n_u^w T_\mathrm{S}\right)}\right]. \end{aligned} \tag{5.2}$$

An example is shown in Table 5.1 shows the ranking of the detected peaks with pairs from Fig. 5.1. The strongest peak has a time lag of 2 hence the start point of the signal is at the sample index of 3. Based on the scenario setup described in Section 4.3.1, the process of the interference cancellation will be shown through the subsequent constellation diagram in this chapter. The constellation diagram of the received signal and the time synchronized received signal containing only one Comm signal (along with the Radar's backscattered signals) is shown in Fig. 5.2(a). It can be seen that besides the phase rotation the symbols also suffer from amplitude attenuation (cf. Fig. 4.5(b)). After the time correction, the training symbols become visible in Fig. 5.2(b). These belong to the first SCA training symbol, where the energy is made to be twice as much as the regular modulation symbols, hence their easy visibility. The outer ring of symbols belong to the Comm while the symbols gathered at $(0,0)$ belong to the weaker desired signals.

Table 5.1: XCORR results based on the outcome in Fig. 5.1.

Peak rank	Start point	Peak amplitude
1	3	0.00023822
2	515	$6.9763 \cdot 10^{-}5$
3	1027	$2.5511 \cdot 10^{-}5$
4	$2.3501 \cdot 10^5$	$2.3103 \cdot 10^{-}5$
5	$2.3527 \cdot 10^5$	$2.2095 \cdot 10^{-}5$
6	$2.1543 \cdot 10^5$	$2.1736 \cdot 10^{-}5$

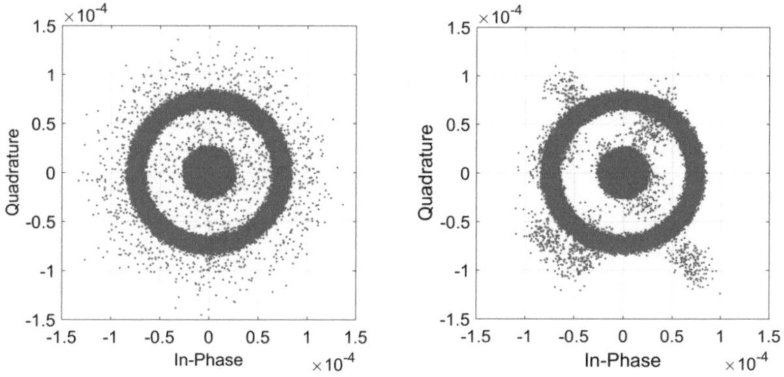

(a) Received symbols before processing. (b) Received symbols after time synchronization.

Fig. 5.2: Received modulation symbols and training symbols (a) before, and (b) after the first time synchronization. After the initial time correction, the SCA training symbols with twice the energy of regular modulation symbols become visible. The amplitude attenuation has not been corrected in this step (cf. Fig. 4.5(b)).

5.2 Frequency synchronization

Using the same SCA training symbols, the first-stage frequency offset can be estimated and corrected. The SCA in essence, estimates the frequency offset along the frequency-subcarrier axis of the OFDM frame. When there are multiple overlapping signals (i.e. from multipaths, and other RadCom nodes), the accuracy of the SCA frequency estimate is very much dependent on the stronger components' values. In the case where the $\mathrm{SIR_{in}}$ is sufficiently large (i.e. when the power difference between the Radar's and Comm's signals is small) or the power difference between the strongest signal and the next strongest signal is small, the SCA tend to run into error. The frequency offset of the multiple overlaid signals will add up as a vector sum, yielding an estimate that does not approach the true value. This will be addressed in the Case Studies section at the end of this chapter. Meanwhile, the Comm's power must be adequately high enough for the SCA to yield a 'coarse' estimate that approaches the real offset value.

Since it is known that the frequency offset estimated by the SCA is based on d^* (see Section 4.2.1), and that d^* is not employable for radar cases, the SCA algorithm is slightly modified to take the value of N_{τ_w} as the input to produce $\tilde{\delta}_w = \frac{\angle P(N_{\tau_w})}{\pi T_0}$ (cf. Eq. (4.17)), where $\tilde{\delta}_w$ is the output of the SCA algorithm. In the case of any residue frequency offset, it can be written as $(\delta_w - \tilde{\delta}_w)$. The first-stage frequency synchronization is then done by multiplying the term $\mathrm{e}^{-\mathrm{j}2\pi(\tilde{\delta}_w)t}$ to the time domain corrected received signal of Eq. (5.2) to result in

$$\begin{aligned}\widetilde{\mathbf{W}}^{\delta}(n_u^w, m) = \mathbf{S}_w(n_u^w, m)\,\alpha_w\,\mathrm{e}^{\mathrm{j}(\Psi_w+\zeta_w)}\left[\mathrm{e}^{\mathrm{j}\varphi_{\mathrm{R}_w}(u^w)}\,\mathrm{e}^{-\mathrm{j}2\pi i N_{\mathrm{ch}}\Delta f(\tau_w-\tilde{\tau}_w)}\right]\cdot \\ \left[\mathrm{e}^{\mathrm{j}\varphi_{\mathrm{train},w}}\,\mathrm{e}^{\mathrm{j}2\pi(\delta_w-\tilde{\delta}_w)\left(mT+n_u^w T_{\mathrm{S}}\right)}\right].\end{aligned} \tag{5.3}$$

Fig. 5.3 shows the SCA corrected modulation symbols in their constellation diagram, (a) with and (b) without the training symbols. From the orientation of the training symbols in Fig. 5.3(a) (cf. original training symbols in Fig. 4.5(b)) it is obvious that there is still about 22° of phase rotation clockwise to be done to put the modulation symbols back to their correct quadrants.

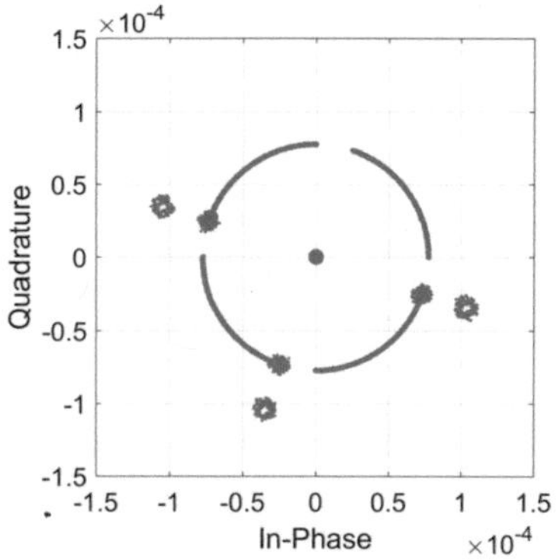

(a) Modulation symbols with training symbols.

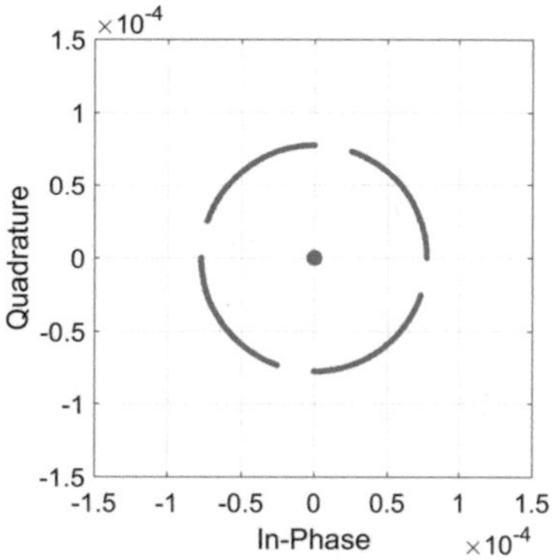

(b) Modulation symbols only.

Fig. 5.3: Received modulation symbols (a) with, and (b) without training symbols, after SCA frequency synchronization. The cluster of points gathered around $(0,0)$ contains the symbols from weaker signals, i.e. the desired Radar's backscattered symbols. The SCA training symbols are generated with a missing upper right corner (cf. Fig. 4.5(b)). At this point, the SCA training symbols shows that there is an approximately 22° clockwise rotation to the correct quadrants. The amplitude attenuation has not been corrected in this step (cf. Fig. 4.5(b)).

5.3 Channel coefficients

Next, the residues from the estimation of the time delay and frequency offset that exceed one bin will be estimated and corrected. This is done with the aid of the pilot tones, which are used to gauge the change of the channel over one transmit frame. This channel state information (CSI) accounts for all the wave propagation effects, time-varying nature of the channel due to the propagation environment, and to a certain extend the hardware imperfections. These changes in the channel will reflect in the pilot tones hence with a comparison with the original pilot tones, these changes become the channel transfer function. Due to the cyclic nature of the OFDM symbols with the CP, the channel transfer function is a simple element-wise division of the received pilot tones and the original pilot tones.

This element-wise operation yields one matrix for every qp-th frame and contains the 'channel coefficients'. This is exactly the same procedure as followed by the radar in estimating the range and Doppler as shown in Eq. (2.41). Hence it is also possible to correct the time delay residue and frequency offset with Fourier transform operations using the channel coefficients. To obtain sufficiently accurate residue time and frequency offset estimation using the channel coefficients, several factors must be considered namely the,

- SNR and $\mathrm{SIR_{in}}$. The higher the Comm signal power compared to the noise floor and the radar signal, the better the estimation.
- power difference between the Comm nodes. If there is more than one Comm node, the power difference between the Comm is analogous to the SNR and $\mathrm{SIR_{in}}$. For a reliable estimate, the power difference must be at least 10 dB.
- first-stage time and frequency synchronization. The first-stage corrections must have corrected a sufficient amount of time shift and frequency offset leaving residues that can be corrected using the pilot tones.
- amount of residue frequency offset $(\delta_w - \tilde{\delta}_w)$ and time delay $(\tau_w - \tilde{\tau}_w)$. The amount that can be corrected is dependent on the spacing of the pilot tones within the OFDM transmit frame, N_t and N_f. These spacings adhere to the Nyquist theorem and dictate the maximum possible amount of correction.

To extract the pilot tones their indices on the frequency-subcarrier and time-symbol axes based on the spacings given by N_f and N_t respectively are first defined. As have been mentioned in the previous chapter, the pilot tones will also line the last row and column of the frame, hence their indices can be written as

$$\begin{aligned} n'_u &= \left\{u,\ u+N_\mathrm{f}N_\mathrm{ch},\ u+2N_\mathrm{f}N_\mathrm{ch}, \ldots,\ u+\left(\frac{N}{N_\mathrm{f}N_\mathrm{ch}}-1\right)N_\mathrm{f}N_\mathrm{ch},\ u+N-N_\mathrm{ch}\right\} \\ &= \left\{u+\mu N_\mathrm{f}N_\mathrm{ch},\ u+N-N_\mathrm{ch}\right\}, \qquad \mu = 0, \ldots, \tfrac{N}{N_\mathrm{f}N_\mathrm{ch}}-1 \\ m' &= \left\{1,\ 1+N_\mathrm{t},\ 1+2N_\mathrm{t}, \ldots,\ 1+\left(\frac{N}{N_\mathrm{t}}-1\right)N_\mathrm{t},\ M\right\} \\ &= \left\{1+\nu N_\mathrm{t},\ M\right\}, \qquad \nu = 0, \ldots, \tfrac{N}{N_\mathrm{t}}-1. \end{aligned} \tag{5.4}$$

The received pilot tones from the Comm $\mathcal{P}_w$, to the Radar's q-th Rx $\mathcal{P}_{qw}$, can be expressed as

$$\mathcal{P}_{qw}(n'_u, m') = \mathbf{H}_w(n'_u, m') \cdot \mathcal{P}_w(n'_u, m') + \mathbf{Z}(n'_u, m'). \tag{5.5}$$

Omitting the noise term $\mathbf{Z}(n'_u, m')$ for simplicity, the estimated channel coefficient matrix using element-wise division of the pilot tones is then,

$$\widetilde{\mathbf{H}}_w(n'_u, m') = \frac{\mathcal{P}_{qw}(n'_u, m')}{\mathcal{P}_w(n'_u, m')}. \tag{5.6}$$

To obtain the channel matrix with the same size as the full frame matrix of $\mathbb{C}^{N \times M}$, a Spline interpolation of $\widetilde{\mathbf{H}}_w(n'_u, m')$ is then done to result in $\widehat{\mathbf{H}}_w^{\text{frame}}(n, m)$ give in Eq. (5.7). For each set of subcarrier n_u^w of the Comm(s) there will be a different $\widehat{\mathbf{H}}_w^{\text{frame}}(n, m)$.

$$\widetilde{\mathbf{H}}_w(n'_u, m') \;\in\; \widehat{\mathbf{H}}_w^{\text{frame}}(n, m), \qquad \text{for } n = 0, ..., N-1 \qquad m = 0, ..., M-1 \tag{5.7}$$

5.3.1 Range residue estimation

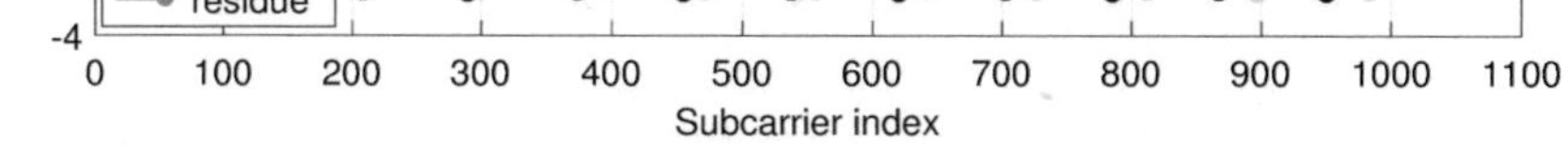

Fig. 5.4: Phase rotation over OFDM subcarrier index due to the true range and the sampled time delay that is correct to the nearest time resolution bin. The difference between the true and sampled time delay is the residue that grows along with the subcarrier index. The simulated target is at $r = 20\,\text{m}$ with $N = 1024$ and $f = 24\,\text{GHz}$. The sampling rate is $T_\text{S} = T_0/N = 10.7\,\text{ns}$.

As discussed in Section 5.1, the time synchronization can only correct up to within one range resolution bin. To show the effect of the phase rotation due to the true range versus the sampled time delay estimated by the XCORR to the nearest time bin, a simple simulation is set up in Matlab. With an operating frequency of 24 GHz, $N = 1024$ and the target distance being $r = 20\,\mathrm{m}$, the phase rotation over subcarriers based on the 'true' and XCORR 'sampled' time delay is as shown in Fig. 5.4. The difference between them is the residue time delay that grows over the subcarrier index.

Based on the sampling theory to determine the pilot tone spacings in Eq. (4.22), the maximum number of residue time lags $N_{\tau_2} \in \mathbb{Z}$, that can be correctly estimated in this step without estimation ambiguity is

$$\begin{aligned} N_{\tau_2} &= \frac{\tau_{\max}}{T_\mathrm{S}} = \frac{1}{2N_\mathrm{f}\Delta f N_\mathrm{ch} T_\mathrm{S}}, \qquad T_\mathrm{S} = \tfrac{1}{N\Delta f} \\ &= \frac{N}{2N_\mathrm{f}N_\mathrm{ch}}. \end{aligned} \tag{5.8}$$

If $N_{\tau_2} > 1$, this integer residue time lag can be found by taking an IDFT of $\widehat{\mathbf{H}}_w^{\mathrm{frame}}(n,m)$ to yield

$$\begin{aligned} \widehat{\mathbf{H}}_\mathbf{T}(k,m) &= \mathrm{IDFT}\left[\widehat{\mathbf{H}}_w^{\mathrm{frame}}(n,m)\right] \\ &= \sum_{n=0}^{N-1} \widehat{\mathbf{H}}_w^{\mathrm{frame}}(n,m) \cdot \mathrm{e}^{\mathrm{j}2\pi\frac{n}{N}k} \\ &= \alpha_w \mathrm{e}^{\mathrm{j}(\Psi_w+\zeta_w)} \cdot \sum_{n=0}^{N-1}\left[\mathrm{e}^{-\mathrm{j}2\pi n\Delta f(\tau_w-\tilde{\tau}_w)}\mathrm{e}^{\mathrm{j}2\pi\frac{n}{N}k}\right]. \end{aligned} \tag{5.9}$$

Of interest here is just the amount of shifted time delay samples given by $\hat{k}$, which can be any integer within the values bounded by $\{0,1,...,N_{\tau_2}\}$. Equating both exponential terms within the square brackets, the maximum will occur at the index $\hat{k}$, where

$$\hat{k} = N\Delta f(\tau_w - \tilde{\tau}_w), \qquad \text{condition}: \; \hat{k} \in \left\{0,1,...,N_{\tau_2}\right\}. \tag{5.10}$$

Let the uncorrected or residue time delay in this step be $\breve{\tau}_w$. After the time delay correction, again by shifting the time domain signal samples according to $\hat{k}$, the Comm modulation symbols can be written as

$$\begin{aligned} \widetilde{\mathbf{W}}^{\tau_2}(n_u^w,m) = \mathbf{S}_w(n_u^w,m)\alpha_w \mathrm{e}^{\mathrm{j}(\Psi_w+\zeta_w)}&\left[\mathrm{e}^{\mathrm{j}\varphi_{\mathrm{R}_w}(u^w)}\mathrm{e}^{-\mathrm{j}2\pi i N_\mathrm{ch}\Delta f\breve{\tau}_w}\right]\cdot \\ &\left[\mathrm{e}^{\mathrm{j}\varphi_{\mathrm{train},w}}\mathrm{e}^{\mathrm{j}2\pi(\delta_w-\tilde{\delta}_w)\left(mT+n_u^w T_\mathrm{S}\right)}\right]. \end{aligned} \tag{5.11}$$

If at this point or during the first-stage XCORR time synchronization the corrected time delay falls within one range bin, no further estimation is needed because the residual phase rotations will be corrected along with other channel effects in the channel equalization matrix.

5.3.2 Fine frequency offset estimation

In the same way as using a DFT over the time-symbol axis of Eq. (2.41) to estimate the Doppler, the same can be done to estimate the residue frequency offset using the interpolated channel coefficient matrix. The intention here is to estimate the residue frequency offset in addition to the one estimated coarsely by the SCA so that the total estimated frequency offset is as near as possible to the real frequency offset. Unlike in the residue range phase rotation, which influences only the frequency-subcarrier axis, a frequency offset affects both the frequency and time axes. Hence the closer the estimation to the real frequency offset, the 'cleaner' the channel estimation matrix will be.

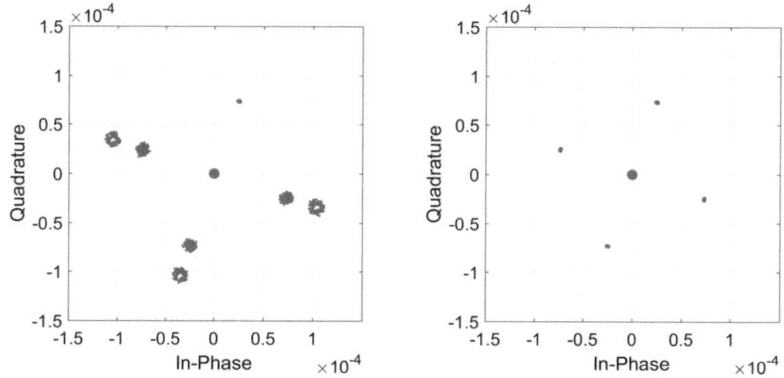

(a) Modulation symbols with training symbols.

(b) Modulation symbols only.

Fig. 5.5: Received modulation symbols and training symbols after fine frequency synchronization step (a) with training symbols, and (b) with only the payload data symbols of Comm. The cluster of points gathered around (0, 0) contains the symbols from weaker signals, i.e. the desired Radar's backscattered signals. The amplitude attenuation has not been corrected in this step (cf. Fig. 4.5(b)).

In order to use this fine frequency estimation to yield a reasonable result, there are two conditions to be met - 1) most of the frequency offset has been corrected by the SCA leaving only residues that are within the capabilities of the pilot tones to correct (refer to Eq. (4.21)) and, 2) zero-padding is used during the DFT process. While zero-padding will not increase the resolution, it can help with the accuracy of the estimated residue frequency offset by pin-pointing the exact bin that the peak falls into. Using a zero-padding factor of $z_{\text{pad}} \in \mathbb{Z}$ (in powers of 2), the resolution of the residue frequency offset estimate is given by

$$\Delta\delta = \frac{1}{M z_{\text{pad}} T}. \tag{5.12}$$

Taking a DFT of the zero-padded channel estimation matrix will yield

$$\begin{aligned} \widehat{\mathbf{H}}_{\mathbf{F}}(n,l) &= \text{DFT}\left[\widehat{\mathbf{H}}_{w}^{\text{frame}}(n,\mu)\right], \qquad \mu = 0,\ldots,M\cdot z_{\text{pad}}-1 \\ &= \sum_{\mu=0}^{M\cdot z_{\text{pad}}-1} \widehat{\mathbf{H}}_{w}^{\text{frame}}(n,\mu)\cdot \mathrm{e}^{-\mathrm{j}2\pi\frac{\mu}{M\cdot z_{\text{pad}}}l} \\ &= \alpha_w \mathrm{e}^{\mathrm{j}(\Psi_w+\zeta_w)}\cdot \sum_{\mu=0}^{M\cdot z_{\text{pad}}-1}\left[\mathrm{e}^{\mathrm{j}2\pi(\delta_w-\tilde{\delta}_w)\mu T}\,\mathrm{e}^{-\mathrm{j}2\pi\frac{\mu}{M\cdot z_{\text{pad}}}l}\right]. \end{aligned} \tag{5.13}$$

Equating both exponential terms within the square brackets, the maximum will occur at index $\hat{l}$, where

$$\hat{l} = M\cdot z_{\text{pad}}(\delta_w-\tilde{\delta}_w)T, \qquad \hat{l}\in\left\{0,1,\ldots,M\cdot z_{\text{pad}}-1\right\}. \tag{5.14}$$

The estimated frequency offset is then $(\delta_w-\tilde{\delta}_w)\approx \hat{l}\Delta\delta$. This estimation is not exact hence there will again be some residue frequency offset. Let the residue frequency offset after this step be $\breve{\delta}_w = \delta_w-\tilde{\delta}_w-\hat{l}\Delta\delta$. After the second-stage frequency correction by means of multiplication of the time domain signal (from the previous step) with $\mathrm{e}^{-\mathrm{j}2\pi\hat{l}\Delta\delta t}$ the Comm's modulation symbols become

$$\begin{aligned} \widetilde{\mathbf{W}}^{\delta_2}(n_u^w,m) = \mathbf{S}_w(n_u^w,m)\,\alpha_w\,\mathrm{e}^{\mathrm{j}(\Psi_w+\zeta_w)}\left[\mathrm{e}^{\mathrm{j}\varphi_{\mathrm{R}_w}(u^w)}\,\mathrm{e}^{-\mathrm{j}2\pi i N_{\text{ch}}\Delta f\,\check{\tau}_w}\right]\cdot \\ \left[\mathrm{e}^{\mathrm{j}\varphi_{\text{train},w}}\,\mathrm{e}^{\mathrm{j}2\pi\breve{\delta}_w(mT+n_u^w T_{\mathrm{S}})}\right]. \end{aligned} \tag{5.15}$$

After this step, if the total phase rotation of the leftover uncorrected time delay and frequency offset is small enough, the cluster of the modulation symbols will be small as shown in Fig. 5.5. It can be seen from Fig. 5.5(a), from the rotation of the training symbols that the time and frequency synchronization has not managed to rotate the modulation symbols back to their correct quadrants.

5.4 Channel equalization matrix

In this final step of correction, any leftover uncorrected time delay and frequency terms, as well as fixed phase terms that cannot be estimated from the aforementioned steps are corrected here. In this step, the amplitude of the Comm signal is also equalized and any random phase terms that affect the individual symbols can also be corrected. Following the steps in Eq. (5.5) to Eq. (5.7), the channel matrix of the second-stage corrected signal is estimated again to yield $\widehat{\mathbf{H}}_w^{\text{frame},2}(n,m)$, whereby

$$\widehat{\mathbf{H}}_w^{\text{frame},2}(n,m) = \alpha_w \mathrm{e}^{\mathrm{j}(\Psi_w+\zeta_w)} \left[\mathrm{e}^{-\mathrm{j}2\pi n\Delta f \check{\tau}_w}\right] \cdot \left[\mathrm{e}^{\mathrm{j}\varphi_{\text{train},w}} \mathrm{e}^{\mathrm{j}2\pi\check{\delta}_w\left(mT+n_u^w T_\mathrm{S}\right)}\right]$$
$$\text{where } n = 0,1,...,N-1 \text{ and } m = -1,-2,0,1,...,M-1. \tag{5.16}$$

Notice here that the number of subsequent OFDM symbols (with the index m) have been increased by two to account also for the training symbols. This then leads to a channel equalization matrix of

$$\widehat{\mathbf{G}}_w(n,m) = \frac{1}{\widehat{\mathbf{H}}_w^{\text{frame},2}(n,m)}. \tag{5.17}$$

When the Comm's modulation symbols are element-wise multiplied with the channel equalization matrix at subcarrier indices of n_u^w (with the help of the Comm's subcarrier mask) this will give an estimate of the Comm's payload data according to

$$\begin{aligned}\widehat{\mathbf{W}}(n_u^w,m)) &= \widetilde{\mathbf{W}}^{\delta_2}(n_u^w,m) \circ \widehat{\mathbf{G}}(n_u^w,m), \qquad m = 0,1,...,M-1 \\ &\approx \mathbf{S}_w(n_u^w,m).\end{aligned} \tag{5.18}$$

If the total residue phase after this stage of correction is small enough for the modulation symbols to be contained within their correct quadrants in the constellation, the demodulated message signal from Comm will contain minimal to no error. At this stage, the received modulation symbols can be written as

$$\widehat{\mathbf{Y}}_q(n,m) = \sum_{p=0}^{P-1} \mathbf{X}_p(n_u,m) \sum_{h=0}^{H-1} \alpha_h \mathrm{e}^{\mathrm{j}(\Psi_{p,q,h}+\zeta_{n,m})} \left(\mathrm{e}^{\mathrm{j}\varphi_{\mathrm{R}_h}(u)}\, \mathrm{e}^{-\mathrm{j}2\pi(iN_{\mathrm{ch}})\Delta f \tau_h}\right) \cdot \left(\mathrm{e}^{\mathrm{j}\varphi_{\mathrm{train}}}\, \mathrm{e}^{\mathrm{j}2\pi f_{\mathrm{D},h}(mT+n_u^w T_{\mathrm{S}})}\right) + \widehat{\mathbf{W}}(n_u^w,m) + \hat{\mathbf{Z}}(n,m). \tag{5.19}$$

It becomes obvious that if $n_u^w \neq n_u$ then the radar's received modulation symbols are free from interferers, having only the channel effects caused by the targets. The modulation symbols after the channel equalization is shown in Fig. 5.6, where it can be seen that all symbols are clustered distinctly in their four constellation points with minimal standard deviation.

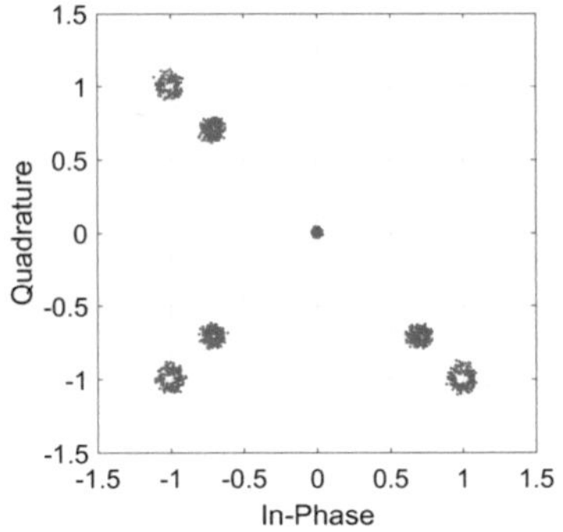

(a) Training symbols only.

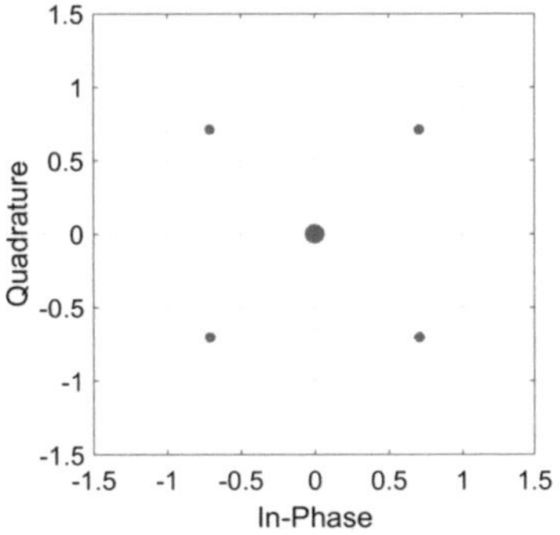

(b) Modulation symbols only.

Fig. 5.6: Received (a) training and (b) modulation symbols, after channel matrix equalization. The Comm's modulation symbols are now located at the four corners of the constellation diagram. The cluster of points gathered around (0,0) contains the symbols from weaker signals, i.e. the desired Radar's backscattered signals. The amplitude attenuation has been corrected in this step (cf. Fig. 4.5(b)).

The recovered training symbols in Fig. 5.6(a) however have a much larger deviation between the points resulting in a bigger cluster size. This is because there are no pilot tones within the training symbols hence the $\widehat{\mathbf{G}}_w(n,m)$ estimation via interpolation from the pilot tones suffers as a result. This however poses no consequence to the recovery of the payload data but contributes slightly to the errors in the reconstructed Comm signal. The desired Radar's modulation symbols of much lower amplitude can be seen as a cluster centering at point $(0,0)$ of the plot.

To recover the message signal, the Comm's symbols are passed through its subcarrier mask to concatenate only the symbols at the subcarriers used by the Comm. The constellation diagrams in Fig. 5.6 will then appear without the cluster of point centered around $(0,0)$. An inverse process of removing the pilot symbols and reversing any additional coding is done then the symbols can be demodulated to produce the message signal. If channel coding (e.g. forward error correction techniques) are used, the demodulation of the signal will have less errors.

5.5 Signal reconstruction and cancellation

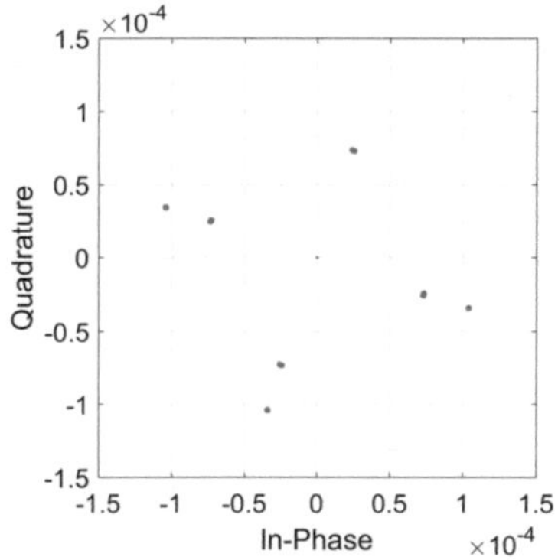

(a) Remodulated symbols with training symbols.

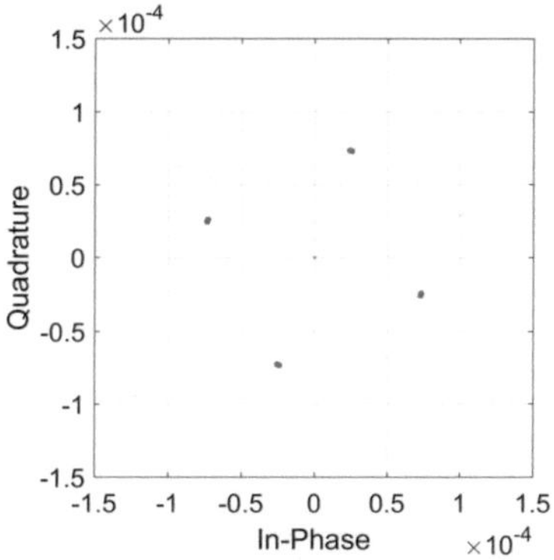

(b) Remodulated symbols only.

Fig. 5.7: Remodulated training and modulation symbols of Comm with channel effects (phase rotation and amplitude attenuation) added (cf. Fig. 5.3).

Using the total estimated time delay sample shift of $N_{\tau_w} + N_{\tau_2}$, total frequency offset of $\tilde{\delta}_w + \hat{l}\Delta\delta$ and the channel equalization matrix $\widehat{\mathbf{G}}_w(n_u^w, m)$, the recovered Comm's training and modulation symbols shown in Fig. 5.6 are added with the aforementioned effects again in the reverse order from the estimation steps described in the previous section. Fig. 5.7 shows the Comm's modulation symbols only with reversed channel equalization, and the remodulated training symbols (assumed known *a priori*). After the addition of the total frequency offset and time delay shifted by the total time sample shift, the constellation diagram in Fig. 5.8 results (cf. Fig. 5.2).

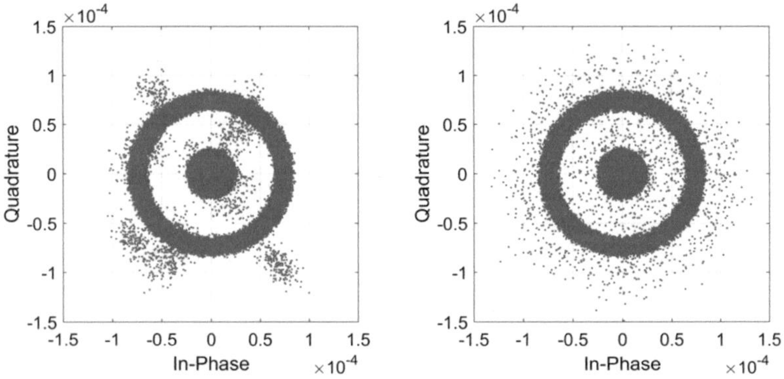

(a) Remodulated symbols with channel effects and frequency offset.

(b) Remodulated symbols with channel effects, frequency offset and time delay.

Fig. 5.8: Remodulated training and modulation symbols of Comm with reversed channel equalization, frequency offset and time delay (cf. Fig. 5.2).

At this point, if the time and frequency estimation as well as the channel equalization matrix are sufficiently accurate, the reconstructed Comm signal will appear approximately as written in Eq. (4.8). This is the Comm signal that is seen at the Radar's Rx, which comprises the payload data and all channel effects. To obtain the Radar's desired signals that are backscattered from the targets, the reconstructed Comm signal is subtracted from the Radar's received time domain signal so that its modulation symbol equivalent $\mathbf{Y}_q(n,m)$ as shown in Eq. (4.13) will be mostly free of the term $\mathbf{W}(n_u^w,m)$ to result in

$$\begin{aligned}\widetilde{\mathbf{Y}}_{q,\text{canceled}}(n,m) = \sum_{p=0}^{P-1} \mathbf{X}_p(n_u,m) \sum_{h=0}^{H-1} \alpha_h \mathrm{e}^{\mathrm{j}(\Psi_{p,q,h}+\zeta_{n,m})} \left(\mathrm{e}^{\mathrm{j}\varphi_{\mathrm{R}_h}(u)}\,\mathrm{e}^{-\mathrm{j}2\pi(iN_{\mathrm{ch}})\Delta f \tau_h}\right)\cdot \\ \left(\mathrm{e}^{\mathrm{j}\varphi_{\mathrm{train}}}\,\mathrm{e}^{\mathrm{j}2\pi f_{\mathrm{D},h}(mT+n_u^w T_{\mathrm{S}})}\right) + \mathbf{W}_{\mathrm{res}}(n_u^w,m) + \hat{\mathbf{Z}}(n,m).\end{aligned} \tag{5.20}$$

where $\widetilde{\mathbf{Y}}_{q,\text{canceled}}(n,m)$ is the interference canceled Radar received frame and $\mathbf{W}_{\text{res}}(n_u^w,m)$ is the residue Comm signal that is not perfectly canceled. The interference canceled radar and DOA images are as shown in Fig. 5.9. Compared to the initial radar and DOA images in Fig. 4.8, all six targets can now be detected without ambiguity and the SNIR_{out} is now 73 dB, an increase of 47 dB. From Eq. (5.19), it can be seen that the angular information of the targets are not affected by the cancellation of the Comm's signal.

If there are more than one Comm or if the Comm is a multi-antenna RadCom node, the interference cancellation algorithm that has been described can be applied in an iterative loop per Rx until a satisfactory result (or 'threshold') is obtained. The flow chart of the looping interference cancellation algorithm is as shown in Fig. 5.10. For a multiple Rx configuration, the interference cancellation flow can be applied in parallel for each Rx. For instance, if the Comm node transmits on four different channels, each of the Radar's Rx simultaneously detects and cancels the Comm's signals in successive loops. A measurement example of a 4-Tx Comm using this technique is presented in Chapter 7.

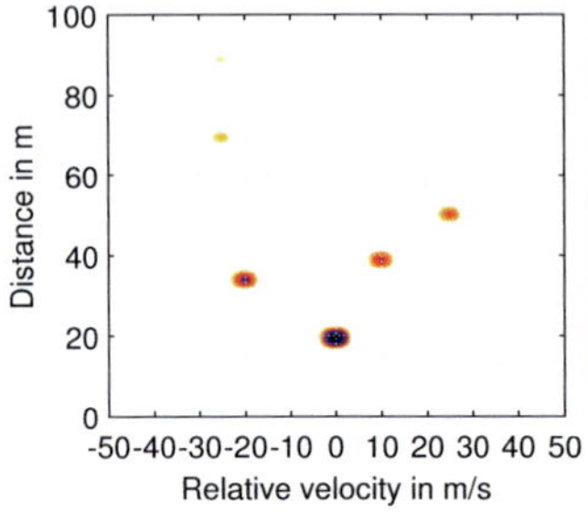

(a) Radar image.

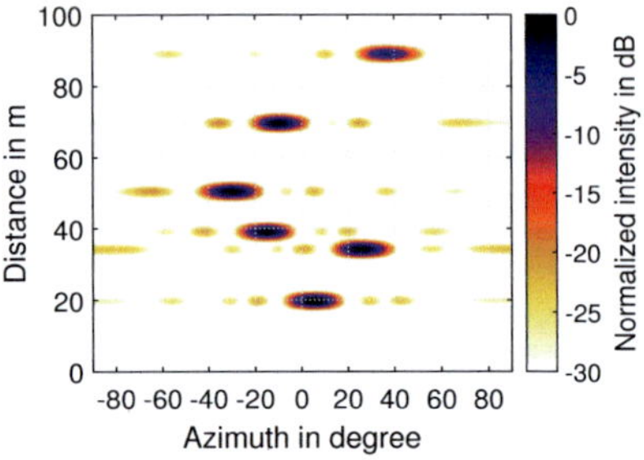

(b) DOA image (summed).

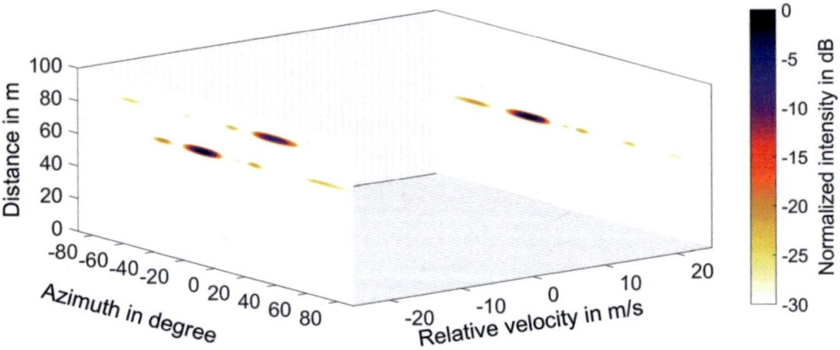

(c) DOA image (slice) of Targets E and F at −25 m/s and D at 25 m/s.

Fig. 5.9: Simulated radar and DOA images after interference cancellation (image before cancellation in Fig. 4.8). Comm signal is at 12.70 m from radar ($\mathrm{SIR_{in}} = -30\,\mathrm{dB}$), $\delta_w T_0 = -0.16$ and SNR = 30 dB. All six targets can now be clearly distinguished in both images without ambiguity. The $\mathrm{SNIR_{out}}$ is 73 dB, an increase of 47 dB from the radar image with interference in Fig. 4.8(a).

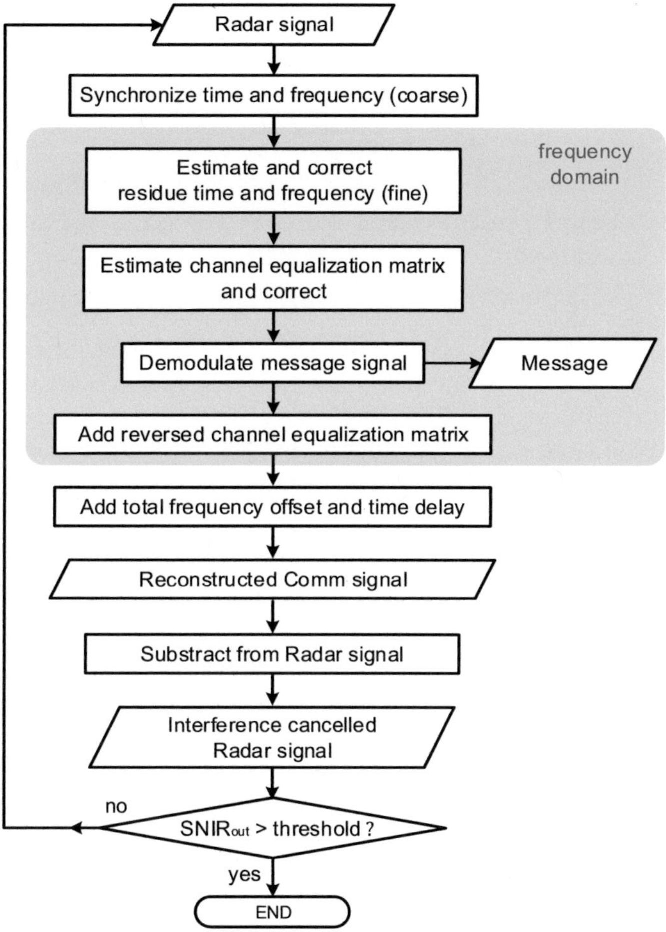

Fig. 5.10: Interference cancellation flow chart. For a multiple Rx configuration, this flow can be applied for each Rx in parallel.

5.6 Case study scenario

Now that the interference algorithm has been presented, some analysis to gauge its performance is in order. Using a similar simulation setup described in Section 4.3.1, with a small modification on the user index assigned to the Radar and Comm(s), three case studies will be presented in the following sections, with:

- Case study I: Spectrally interleaved OFDM signal model with a 1-Tx interferer
- Case study II: Spectrally interleaved OFDM signal model with a 4-Tx interferer
- Case study III: Classical OFDM signal model with a 1-Tx interferer

The performance will be analyzed in terms of the: 1) $\mathrm{SNIR_{out}}$ over $\mathrm{SIR_{in}}$ and frequency offset, and 2) time and frequency synchronization accuracy, 3) Bit error ratio (BER), which is the amount of error of the payload data bits received from Comm compared to the original bits. The BER is given in fractional numbers from 0 to 1 with 0 indicating no error.

Simulation setup for case studies

For a fair comparison over the three case studies, the scenario of six targets with different ranges, velocities and positions (in azimuth angles) as given in Table 4.2 is used along with the OFDM parameters as given in Table 5.2. When no Comm (interferers) are present or when there is no ICI, the maximum achievable $\mathrm{SNIR_{out}}$ for this six-target scenario is around 77 dB.

The Comm and the Radar use different sets of SCA training symbols. Each transmit frame of the same node will have the same preambles. The pilot symbols are the same for all OFDM frames and nodes. It is assumed that the Radar knows *a priori* the preambles and user indices used by the Comm. The message signal (payload data) used for the Comms is a *lorem ipsum* text with each character having an 8-bit ASCII encoding, which fills more than 90% of the available space in the OFDM frame excluding the pilot tones. The rest of the unused space is filled with random symbols within the modulation alphabet. A 4-PSK modulation scheme is used. The analysis will take place at one of the receivers of the Radar. To avoid any bias or unpredictable elements in the

Table 5.2: Radar parameters for the Case Studies I, II and III.

Symbol	Parameter	Value
N	Number of subcarriers that makes up 1 OFDM symbol	1024
M	Number of contiguous OFDM symbols	128
N_{CP}	Number guard interval/cyclic prefix symbols	$N/8$
Δf	Subcarrier spacing	90.909 kHz
$T_0 = 1/\Delta f$	OFDM symbol duration	11 µs
T_{CP}	Duration of guard interval/ cyclic prefix	1.375 µs
$T = T_0 + T_{\mathrm{CP}}$	Total duration of 1 OFDM symbol	12.375 µs
$BW = N \times \Delta f$	Total bandwidth	93.1 MHz
f_c	Carrier frequency	24 GHz
P	Number of transmit antennas with $p = 0, 2, ..., P-1$	4
Q	Number of receive antennas with $q = 0, 2, ..., Q-1$	4
N_{ch}	Number of user channels with $0 \leq N_{\mathrm{ch}} \leq N-1$	8
$N_{\mathrm{f}}, N_{\mathrm{t}}$	Pilot tone spacings in the frequency and time axes	8
Δr	Range resolution	1.61 m

analysis, no noise is added to the simulations and no channel coding is used on the payload data.

There are altogether $N_{\mathrm{ch}} = 8$ channels in the signal bandwidth, with each user index allocated 128 subchannels. The Radar and Comm node are set to transmit the same power. To vary the Comm signals' power to obtain the $\mathrm{SIR}_{\mathrm{in}}$ values, the distance of the Comm from the Radar is varied. The Comm's frequency offset is also varied to obtain the analysis results. In all three case studies, the frequency offsets of $\delta_w T_0 = \{0, -0.03, -0.07, -0.15, -0.3\}$, denoted by f0, f3, f7, f15, f30 in the plots, are applied. This is because shifting the Comm's subcarriers in the negative direction toward the subchannels occupied by the Radar will give a more adverse effect to the Radar's $\mathrm{SNIR}_{\mathrm{out}}$, though after numerous observations from the simulations, this effect is not much greater than when $\delta_w T_0$ is shifted to the positive values. This is in accordance to the effect of CFO presented in Section 4.1.3, where the influence of either a positive or negative frequency shift affects the demodulation of the desired symbols by the same amount.

5.7 Case study I: Interleaved OFDM signal with 1-Tx interferer

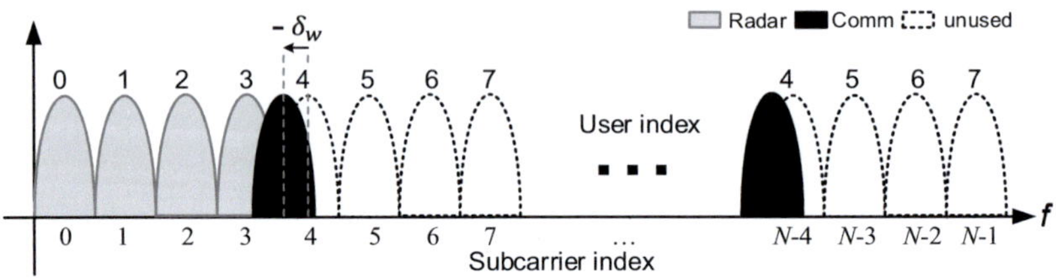

Fig. 5.11: Case Study I: Subcarrier assignment for the Radar $u = \{0,1,2,3\}$ and Comm $u = 4$.

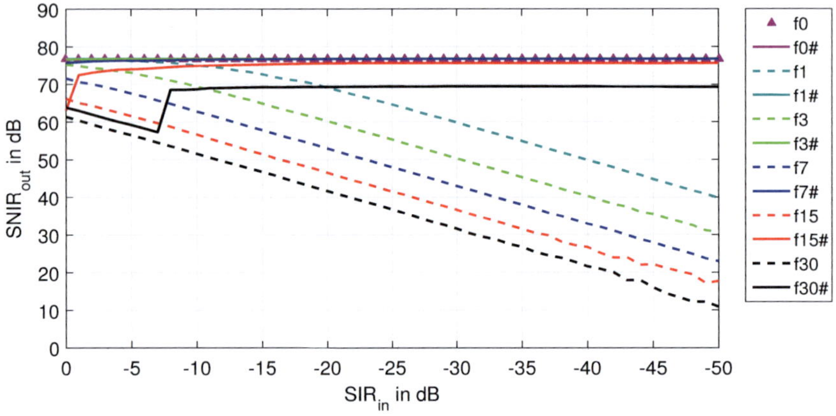

Fig. 5.12: Case Study I: Simulated SNIR_{out} vs. SIR_{in} for one interfering communication partner (Comm) node with one Tx at various frequency offsets. The frequency offsets of $\delta_w T_0 = \{0, -0.03, -0.07, -0.15, -0.3\}$ are denoted by f0, f3, f7, f15, f30. Legends marked with '#' denote the interference canceled equivalent.

This case study starts off with the effect of having one Comm node with one Tx interfering on the Radar. The Radar occupies $u = \{0,1,2,3\}$ (transmitted over different antennas) while Comm occupies only $u = 4$ and the other subcarriers are not used, as depicted in Fig. 5.11. The direction of the frequency offset is indicated by the arrow toward the left side i.e. toward the subchannels occupied by the Radar.

The resulting $SNIR_{out}$ versus SIR_{in} and frequency offset plot is as shown in Fig. 5.12. It can be seen that when there is no frequency offset (f0) the $SNIR_{out}$ remains at the maximum value of around 77 dB with no influence from the decreasing SIR_{in} values (increasing Comm's power or distance). This verifies that so long as the subcarriers remain orthogonal, the isolation between channels is maximum, as stipulated in [SSBZ13]. For the other frequency offset values, before interference cancellation, the $SNIR_{out}$ will decrease linearly with decreasing SIR_{in} till the minimum $SNIR_{out}$ of around 10 dB is met. This value comes from the fluctuations of the noise floor explained in Section 4.3. When the interference cancellation algorithm is applied (legends marked with '#'), two things become obvious. First, as the amount of frequency offset increases, the maximum achievable $SNIR_{out}$ after interference cancellation also becomes lower. Second, with increasing frequency offset, the $SNIR_{out}$ improvement is not apparent until certain SIR_{in} level is reached. These effects can be explained through the time and frequency synchronization accuracy plots.

Fig. 5.13(a) and Fig. 5.13(b) show the time and frequency synchronization accuracy of the interference cancellation algorithm. It can be seen that due to the aid of the preamble, the time synchronization algorithm (first and second stage combined) can detect the start point of the Comm signal accurately. This is evident by all the curves of f1, f3, f7, f15 and f30 ('total') overlapping on the theoretical time delay shift ('theory'). Each delay shift, or time bin here is equivalent to the radar's range resolution. Hence at $SIR_{in} = 0$ dB, the distance of the Comm from the Radar is around 205 m, indicated by the time delay index shift of 127 lags in Fig. 5.13(a).

Estimating the frequency offset however proved to be more challenging. When the frequency offset is adequately small i.e. f1 and f3, the SCA with the help of the channel coefficient estimation of the fine frequency offset ('fine') are able to correctly estimate the total frequency offset. As this frequency offset becomes larger however, especially at the SIR_{in} region of 0 to 7 dB, the Radar's receive signal is more influenced by the Radar's reflected signal power so the SCA fails to provide a good enough coarse frequency estimate. This in turn leaves the channel coefficient matrix still significantly distorted that even the fine frequency offset estimation cannot yield a good estimate. It can be seen that the Comm power must be at least 8 dB higher than the Radar's before the correct f30 estimate is yielded as shown in Fig. 5.13(b). Hence it is also at this SIR_{in} point that the BER drops to almost 0 as shown in Fig. 5.13(c).

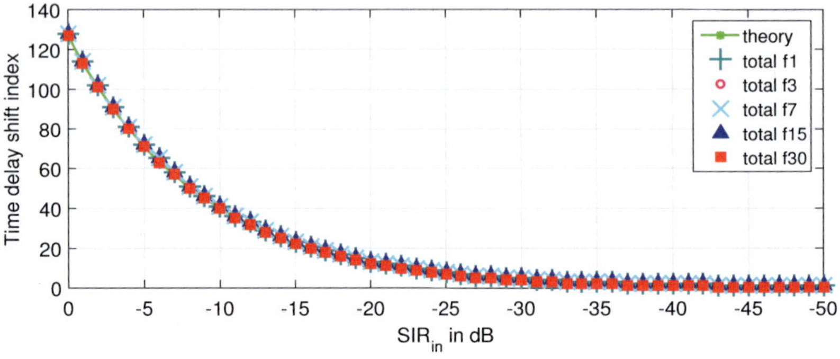

(a) Time synchronization accuracy over SIR_{in} for the detected Comm signal, consisting of the theoretical time delay shift (theory) and the total time delay shift (total).

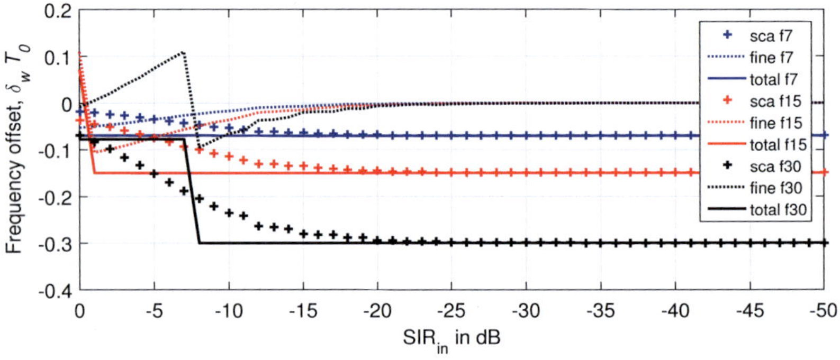

(b) Frequency synchronization accuracy over SIR_{in} for the detected communication signal with $\delta_w T_0 = \{-0.07, -0.15, -0.3\}$, consisting of the SCA outcome (sca), channel coefficient estimation (fine), to yield the total frequency offset (total).

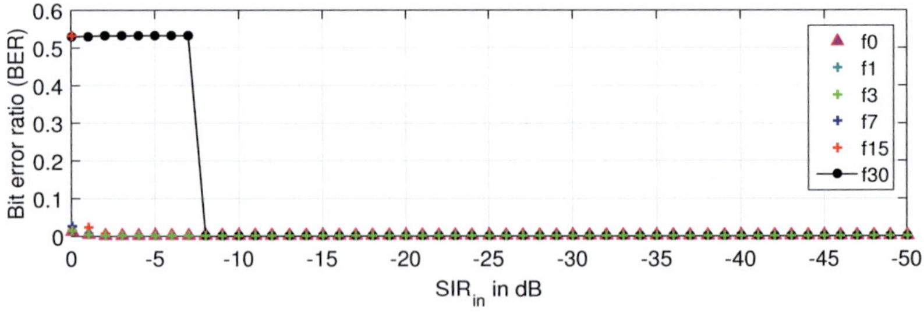

(c) Bit error ratio of the recovered Comm message signal.

Fig. 5.13: Case Study I: Simulated time and frequency synchronization accuracy over $SIR_{in} \in [-50, 0]$ dB, as well as the BER of the recovered Comm message.

The bottomline of this analysis is that the frequency estimation inaccuracy affects the maximum achievable $SNIR_{out}$ and consequently the message recovery in a severe manner. A frequency offset can be viewed as a 'leak' that allows energy from all neighboring channels within the signal bandwidth to contaminate the channel-under-observation. The larger the frequency offset, the larger the leakage and this cannot be fully estimated and resolved even with the best state-of-the-art OFDM frequency synchronization algorithm if the signal in the channel-under-observation is not strong enough. Hence a frequency offset mitigation measure is often implemented beforehand so that any imperfections that lead the loss of subcarrier orthogonality is bounded within a manageable value. In this case study, where only one interferer is present, the message signal can be recovered without error provided that the SIR_{in} is low enough (or the Comm's power is high enough) especially at large frequency offsets. In a scenario with low SNR at the Rx contributed by low transmit power, multipaths propagations and low receiver fidelity, an f30 offset might render all target estimations unreliable even when interference cancellation has been applied.

The radar image and DOA image featuring the interference of $SIR_{in} = -30$ dB with $\delta_w T_0 = -0.16$ and SNR = 30 dB are as shown in Fig. 5.14(a) with the DOA image slices in Fig. 5.14(c). The $SNIR_{out}$ at $u = 0$ is around 36 dB and 26 dB at $u = 3$. This shows that the subchannel that is farthest away from the Comm's channel has the best isolation while the directly neighboring subchannel to the Comm's has the worst isolation. When the color bar of the radar image is scaled to a smaller range, the same image as Fig. 4.8(a) is obtained, whereby only three of the targets nearest to the Radar can be distinguished. The DOA image on the other hand is noisy and cannot be reliably distinguished for some targets.

Applying the interference cancellation algorithm, the interference canceled radar image and DOA image are as shown in Fig. 5.14(b) with the DOA image slices are as shown in Fig. 5.14(d). The $SNIR_{out}$ at $u = 0$ is now 74 dB with the $SNIR_{out}$ at $u = 3$ being 69 dB. The BER is 0. The improvement to the DOA image slices can be observed in Fig. 5.15.

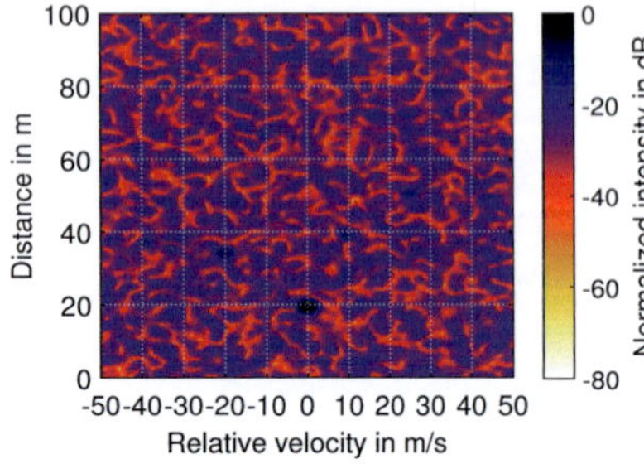

(a) Radar image with interference. $SNIR_{out} = 26\,dB$ at $u = 3$.

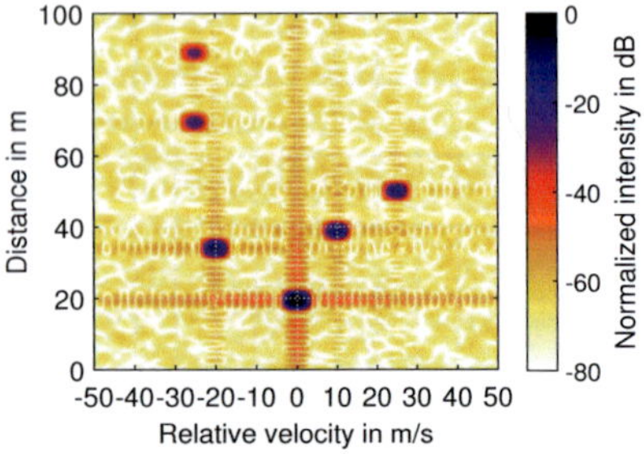

(b) Radar image after interference cancellation. $SNIR_{out} = 69\,dB$ at $u = 3$.

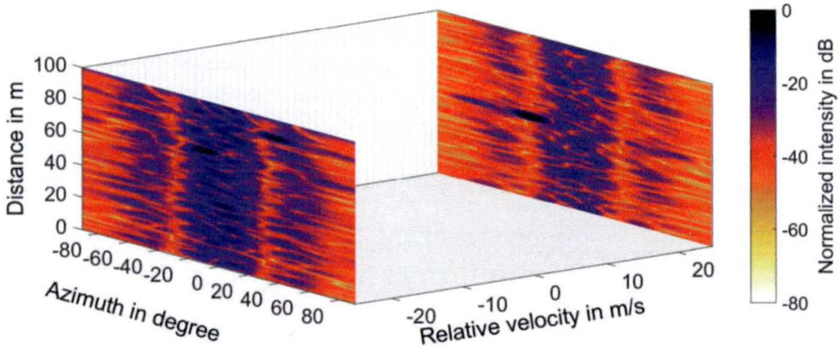

(c) DOA image (slice) with interference. Targets E and F are at −25 m/s and D at 25 m/s.

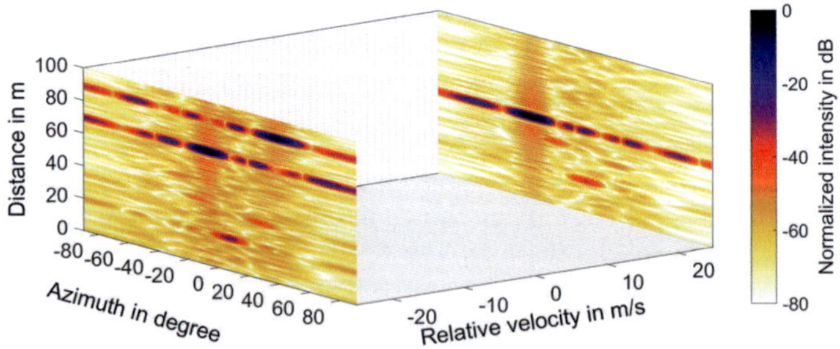

(d) DOA image (slice) after interference cancellation. Targets E and F are at −25 m/s and D at 25 m/s.

Fig. 5.14: Case Study I: Simulated radar image and DOA image slices for $SIR_{in} = -30\,dB$, $\delta_w T_0 = -0.16$ and $SNR = 30\,dB$ (same conditions as used for Fig. 4.8) before and after interference cancellation.

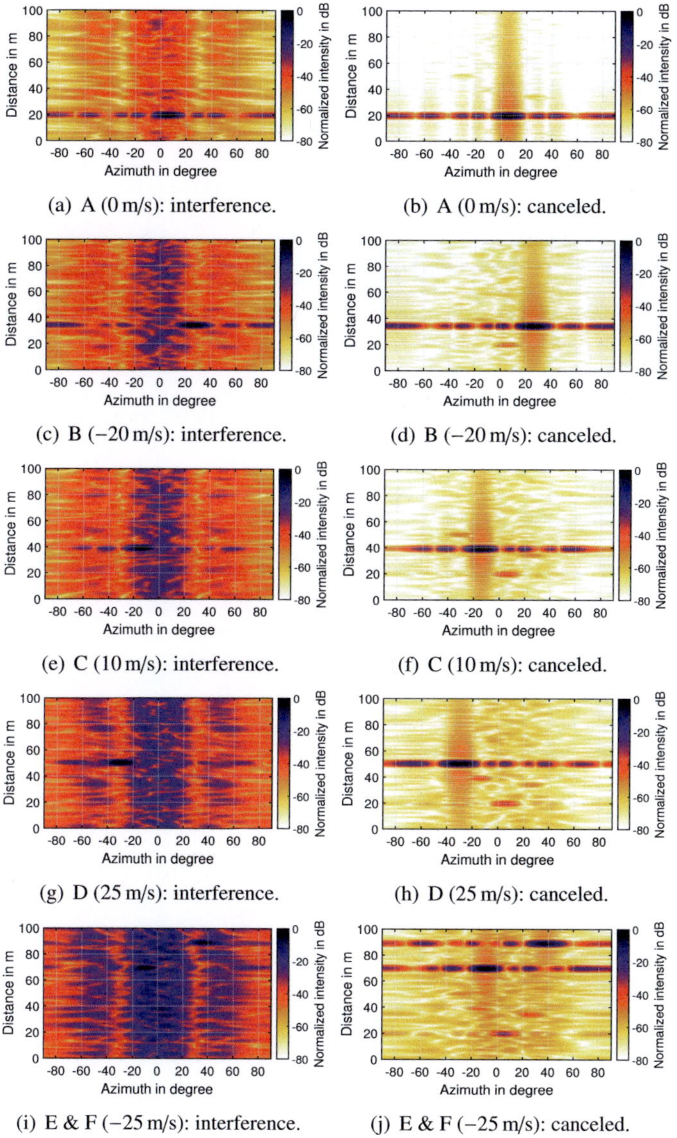

(a) A (0 m/s): interference.
(b) A (0 m/s): canceled.
(c) B (−20 m/s): interference.
(d) B (−20 m/s): canceled.
(e) C (10 m/s): interference.
(f) C (10 m/s): canceled.
(g) D (25 m/s): interference.
(h) D (25 m/s): canceled.
(i) E & F (−25 m/s): interference.
(j) E & F (−25 m/s): canceled.

Fig. 5.15: Case Study I: Comparison of simulated DOA image slices based on velocity according to the scenario given in Fig. 5.14.

5.8 Case study II: Interleaved OFDM signal with 4-Tx interferer

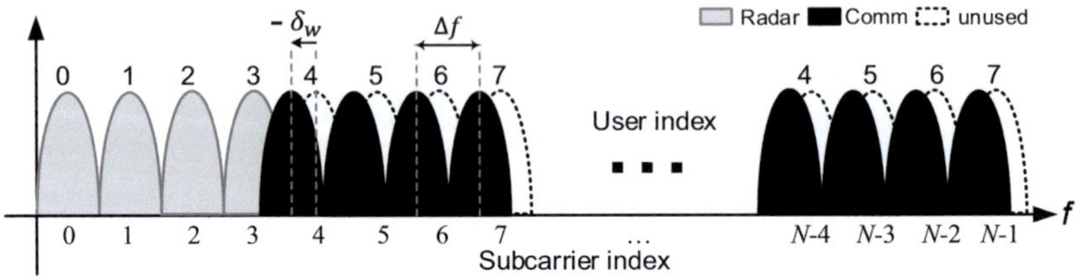

Fig. 5.16: Case Study II: Subcarrier assignment for the Radar $u = \{0,1,2,3\}$ and Comm $u = \{4,5,6,7\}$.

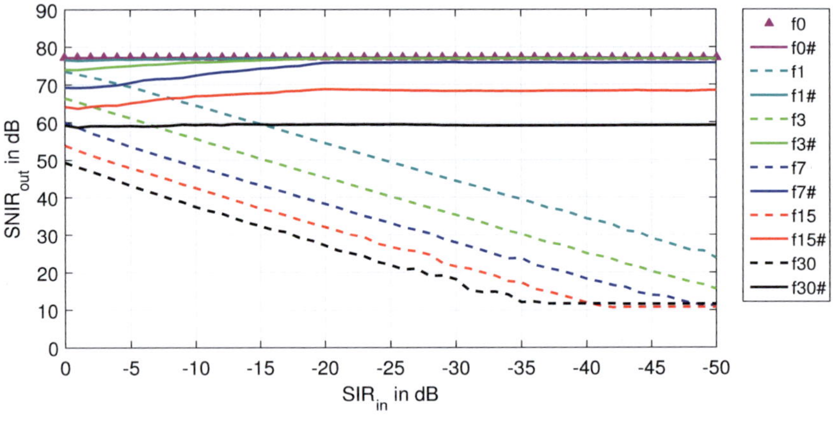

Fig. 5.17: Case Study II: Simulated $\mathrm{SNIR_{out}}$ vs. $\mathrm{SIR_{in}}$ for a 4-Tx interfering communication partner (Comm) node at various frequency offsets. The frequency offsets of $\delta_w T_0 = \{0, -0.03, -0.07, -0.15, -0.3\}$ are denoted by f0, f3, f7, f15, f30. Legends marked with '#' denote the interference canceled equivalent.

In this case study, the interference from a MIMO Comm node with four Txs is of interest. The Radar occupies $u = \{0,1,2,3\}$ while the Comm occupies $u = \{4,5,6,7\}$ as a 4-Tx node as depicted in Fig. 5.16. As a MIMO node, it is assumed that any frequency offset experienced by the Comm node affects all of its channels by the same amount. The propagation path lengths from each of the Comm's Tx to the Radar's Rx are different, but will still be contained within one range bin.

The $SNIR_{out}$ versus SIR_{in} and frequency offset chart is as shown in Fig. 5.17. Similar to Case Study I, the $SNIR_{out}$ decreases linearly with decreasing SIR_{in} for all values of frequency offsets. At the same SIR_{in} point, the $SNIR_{out}$ is lower for the signal with a higher frequency offset. Compared to Case Study I however, the $SNIR_{out}$ for all values of frequency offsets over SIR_{in} are lower by around 10 dB. As a result, the saturation at the $SNIR_{out}$ = 10 dB level also occurs faster, evident from the curves of f15 and f30. In the same way, after the interference cancellation algorithm is applied, the achievable $SNIR_{out}$ is also about 10 dB lower than their counterparts in Case Study I. This effect is especially clear for the f15 # and f30 # curves when comparing Fig. 5.17 with Fig. 5.12. The main difference between the interference canceled $SNIR_{out}$ of Case study I and II is the high SIR_{in} region typically between 0 dB to −10 dB. Unlike in Case Study I, the $SNIR_{out}$ of the interference canceled signals do not decrease but remain constant in this region.

Looking at the time synchronization in Fig. 5.18(a), the start point of the signal over all SIR_{in} and frequency offset values are again correctly determined by the time synchronization algorithm. As for the frequency synchronization accuracy shown in Fig. 5.18(b), all frequency offsets were correctly synchronized over all values of frequency offsets and SIRs. This can be attributed to the Comm signal power, which is now four times larger than in Case Study I (computed on a per channel basis). The Radar signal power is now sufficiently lower than the Comm's and this helps the SCA to estimate the frequency offset with little ambiguity, while the fine frequency estimation makes up the difference of the estimate and the true value. As such there is little BER as shown in Fig. 5.18(c) except for the $SIR_{in} < 4$ dB since the energy of the symbols are not significant enough to for the demodulator to distinguish them over the Radar's symbols correctly.

The radar image and DOA image with the same scenario of $SIR_{in} = -30$ dB, $\delta_w T_0 = -0.16$ and SNR = 30 dB are shown in Fig. 5.19(a) and Fig 5.19(c). The maximum $SNIR_{out}$ of the signal with interference is around 23 dB at $u = 1$ while at $u = 3$ the $SNIR_{out}$ is 20 dB. Since the subchannels of $u = 1$ are directly adjacent to those of $u = 7$ it is not longer the best isolated against ICI but still has a higher $SNIR_{out}$ than $u = 3$ due to the direction of the frequency offset. The subchannels with the highest isolation are $u = \{1,2\}$, both two channels away from the Comm's. The interference canceled radar and DOA image are as shown in Fig. 5.19(b) and Fig. 5.19(c). The maximum $\mathrm{SNIR}_{\mathrm{out}}$ is 69 dB at $u = 1$ while the $\mathrm{SNIR}_{\mathrm{out}}$ at $u = 3$ is 66 dB. Even with the maximum $\mathrm{SNIR}_{\mathrm{out}}$ at around 10 dB lower than in Case Study I, the DOA estimate is still correct with the furthest target at 90 m correctly shown. The improvement to the DOA image slices can be observed in Fig. 5.20.

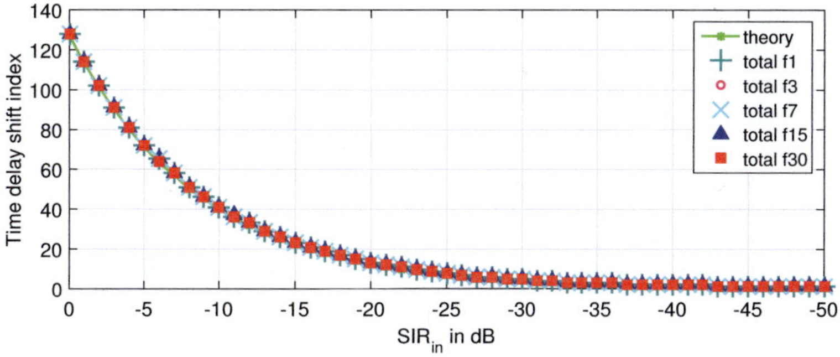

(a) Time synchronization accuracy over $\mathrm{SIR_{in}}$ for the detected Comm signal, consisting of the theoretical time delay shift (theory), cross-correlation outcome (xcorr), channel coefficient estimation (fine), to yield the total time delay shift (total).

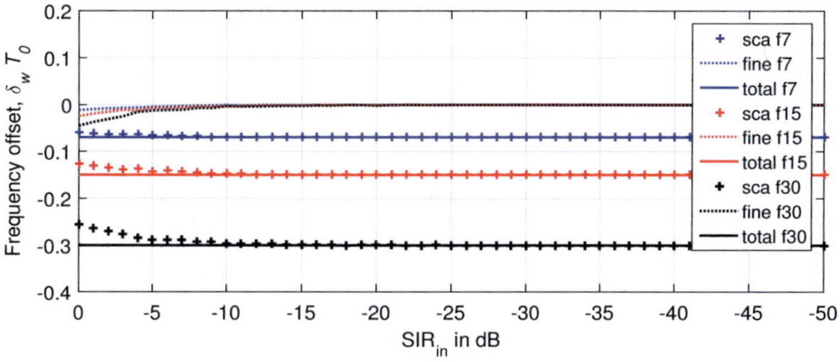

(b) Frequency synchronization accuracy over $\mathrm{SIR_{in}}$ for the detected communication signal with $\delta_w T_0 = \{-0.07, -0.15, -0.3\}$, consisting of the SCA outcome (sca), channel coefficient estimation (fine), to yield the total frequency offset (total).

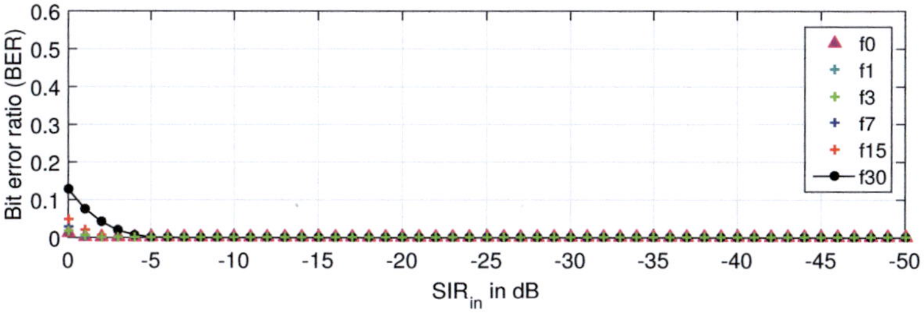

(c) Bit error ratio of the recovered Comm message signal.

Fig. 5.18: Case Study II: Simulated time and frequency synchronization accuracy over $\mathrm{SIR_{in}} \in [-50, 0]$ dB as well as the BER of the recovered Comm message.

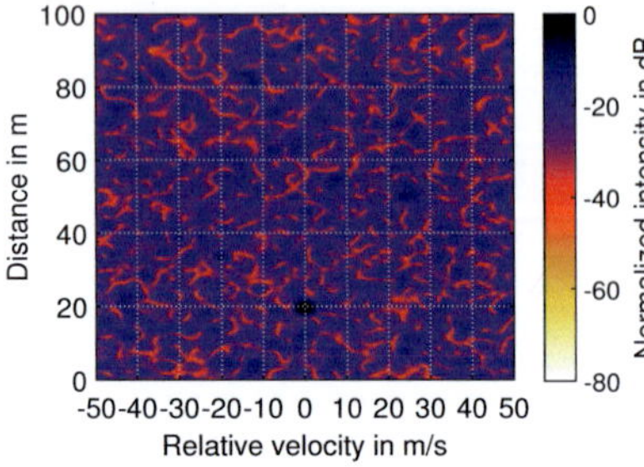

(a) Radar image with interference. $\mathrm{SNIR}_{\mathrm{out}} = 20\,\mathrm{dB}$ at $u = 3$.

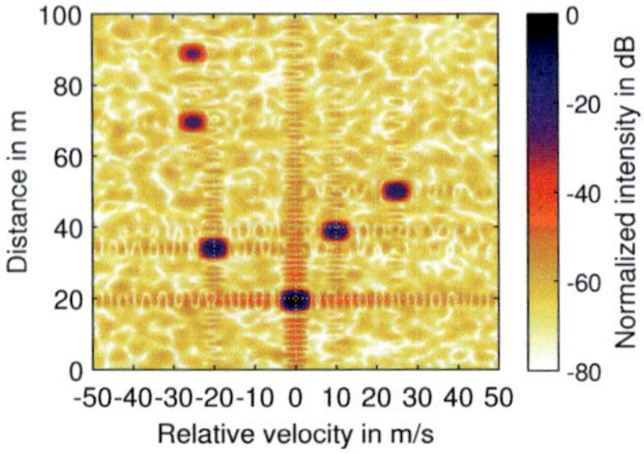

(b) Radar image after interference cancellation. $\mathrm{SNIR}_{\mathrm{out}} = 66\,\mathrm{dB}$ at $u = 3$.

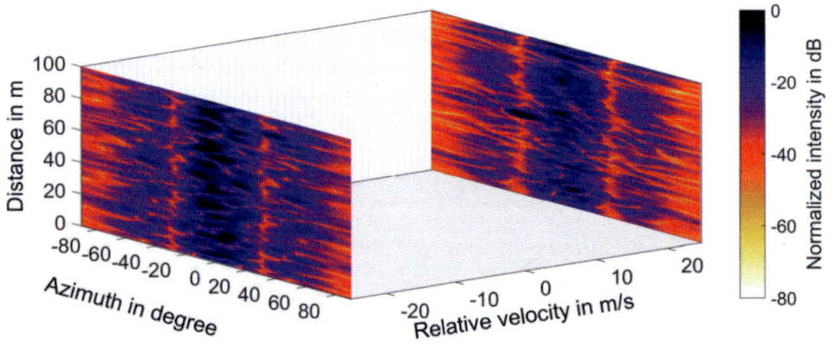

(c) DOA image (slice) with interference. Targets E and F are at −25 m/s and D at 25 m/s.

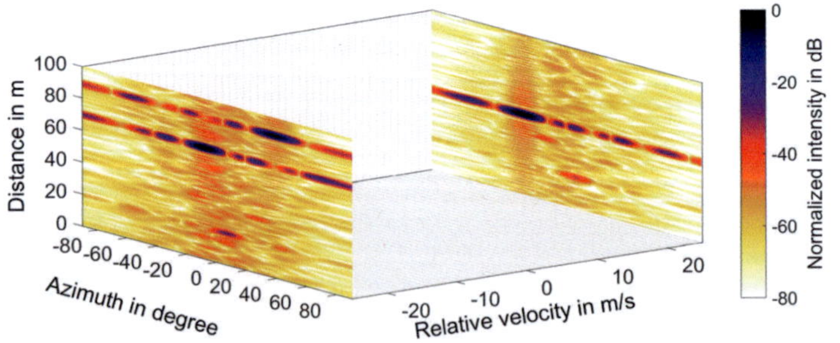

(d) DOA image (slice) after interference cancellation. Targets E and F are at −25 m/s and D at 25 m/s.

Fig. 5.19: Case Study II: Simulated radar image and DOA image for $\mathrm{SIR}_{\mathrm{in}} = -30\,\mathrm{dB}$, $\delta_w T_0 = -0.16$ and $\mathrm{SNR} = 30\,\mathrm{dB}$ (same conditions as used for Fig. 4.8) before and after interference cancellation.

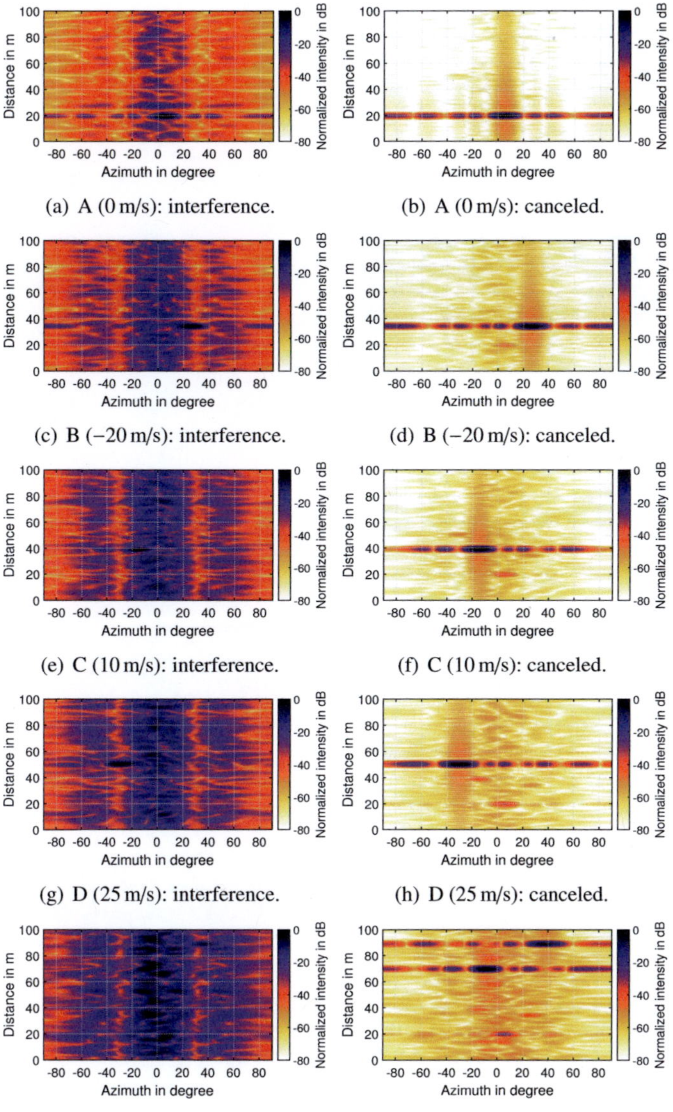

(a) A (0 m/s): interference.

(b) A (0 m/s): canceled.

(c) B (−20 m/s): interference.

(d) B (−20 m/s): canceled.

(e) C (10 m/s): interference.

(f) C (10 m/s): canceled.

(g) D (25 m/s): interference.

(h) D (25 m/s): canceled.

(i) E & F (−25 m/s): interference.

(j) E & F (−25 m/s): canceled.

Fig. 5.20: Case Study II: Comparison of simulated DOA image slices based on velocity according to the scenario given in Fig. 5.19.

5.9 Case study III: Classical OFDM signal with 1-Tx interferer

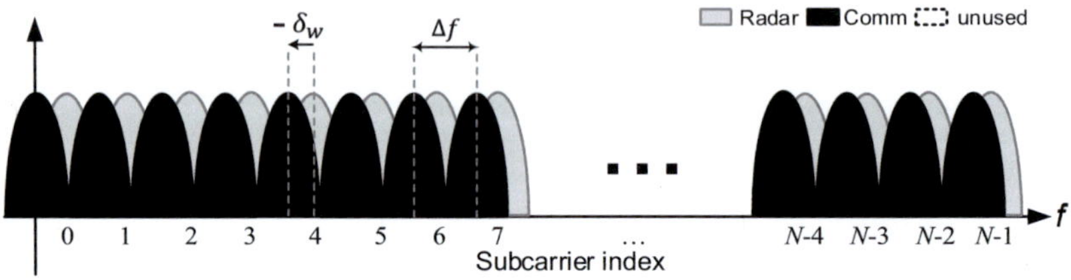

Fig. 5.21: Case Study III: Subcarrier assignment for the Radar $u = 0$ and Comm $u = 0$.

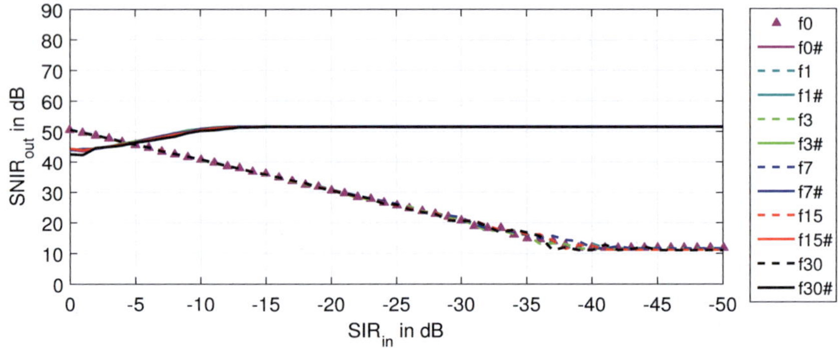

Fig. 5.22: Case Study III: Simulated $\mathrm{SNIR}_{\mathrm{out}}$ vs. $\mathrm{SIR}_{\mathrm{in}}$ for one interfering communication partner (Comm) node at various frequency offsets of $\delta_w T_0 = \{0, -0.03, -0.07, -0.15, -0.3\}$ denoted by f0, f3, f7, f15, f30. Legends marked with '#' signifies the interference canceled equivalent.

This final case study serves to provide a comparison between the robustness of the spectrally interleaved OFDM signal model and the classical OFDM model. Here, the number of channels is set to $N_{\mathrm{ch}} = 1$, inferring that the Radar and Comm both use all of the available N subcarriers within the signal bandwidth for any number of transmit antennas they posses as depicted in Fig. 5.21. To simulate the DOA estimation for a fair comparison with the configuration in Case Study I and II, the Rx of this case is expanded to 16 to produce the same azimuth resolution. To simplify the simulation, both the Radar and Comm is set to use only one Tx. The analysis is then made at one of the Rx of the Radar.

From the SNIR_{out} versus SIR_{in} and frequency offset plot shown in Fig. 5.22, the SNIR_{out} with interference decreases linearly over decreasing SIR_{in} but is the same value over all frequency offset values. This is akin to the f30 curve of Case Study II in Fig. 5.17 (with interference, dashed line). After the interference cancellation is applied to the signal, the resulting SNIR_{out} dips below the SNIR_{out} of the signal with interference and only begin to improve at $\text{SIR}_{\text{in}} = -5\,\text{dB}$ to a maximum of about 51 dB. This result is the same for all frequency offsets of f0 to f30. Even when there is no frequency offset (f0), there is no channel isolation at all.

The time synchronization accuracy is shown in Fig. 5.23(a) and again shows that the time synchronization algorithm has no trouble with distinguishing the start of the signal. Although it is not shown here, the XCORR in the time synchronization starts detecting the strongest peak at the second half of the SCA's first preamble when the frequency offset is around $\delta_w T_0 = \pm 0.35$. In this setup, the estimated time lag puts the start point at more than ±500 lags away (equivalent to a distance of 805 m), which is irrational. The time synchronization algorithm then takes the second strongest peak detected, which is the correct time lag to put through to the SCA. In this way, the reading over all frequency offset values from $\delta_w T_0 \in [-1, 1]$ still fit the curve shown in Fig. 5.22.

The SCA algorithm however, requires Comm's power to be at least 3 dB higher than the Radar's power for an accurate estimation, especially when the frequency offset is high i.e. f30 as can be seen in Fig. 5.23(b). For the lower frequency offset values however, there is no problem with the SCA's estimation. Due to the Comm and Radar sharing the same channel, the channel equalization matrix is distorted especially when the Comm's power is not significant enough in the SIR_{in} of −10 to 0 dB region.

The correct demodulation of the Comm's message signal is only possible when the Comm's power is at least 10 dB higher than the Radar's for this case as shown in Fig. 5.23(c). Although the message can be correctly decoded at this SIR_{in} point, the Radar's SNIR_{out} still suffers at a mere 51 dB. In the presence of more than one Comm Tx (like in Case Study II) or node, the SNIR_{out} as well as the BER of the recovered messages is foreseen to drop even further. When a Comm node with two Tx was tested, the interference canceled result was $\mathrm{SNIR}_{\mathrm{out}} = 20\,\mathrm{dB}$. This shows that such a subcarrier assignment is overly vulnerable toward communication partners (and interference).

Following the same simulation scenario of one Comm node with one Tx that exerts an $\mathrm{SIR}_{\mathrm{in}} = -30\,\mathrm{dB}$ on the Radar's signal with $\delta_w T_0 = -0.16$ and SNR = 30 dB, the resulting radar image is as shown in Fig. 5.14(a) whereby the $\mathrm{SNIR}_{\mathrm{out}}$ is 20.5 dB, comparable to Case Study II. The interference canceled radar image is as shown in Fig. 5.14(b) with the $\mathrm{SNIR}_{\mathrm{out}}$ of 51.6 dB. The DOA image generated from the 16 Radar Rx before and after interference cancellation are as shown in Fig. 5.24(c) and Fig. 5.24(d) respectively. After the interference cancellation, the resulting DOA images still contain traces from Comm rendering the two farthest targets indistinguishable from the ghost targets caused by Comm's residue signals and can be observed in Fig. 5.25.

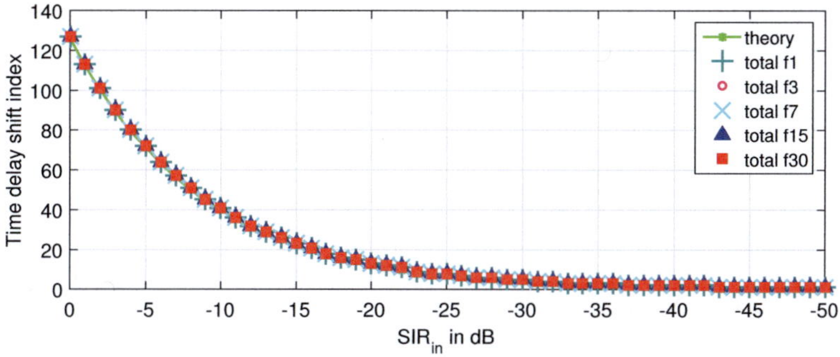

(a) Time synchronization accuracy over SIR_{in} for the detected Comm signal, consisting of the theoretical time delay shift (theory), cross-correlation outcome (xcorr), channel coefficient estimation (fine), to yield the total time delay shift (total).

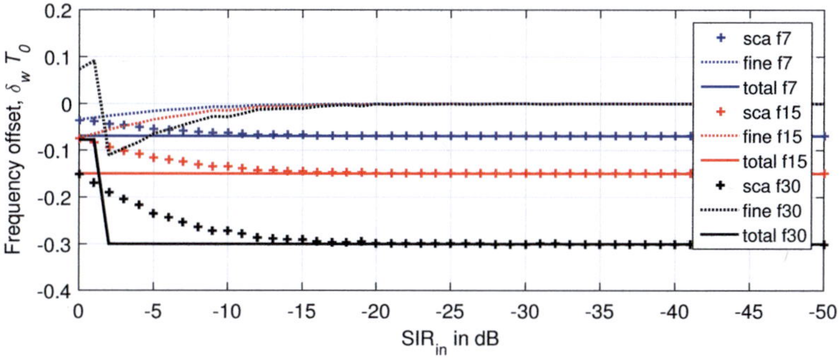

(b) Frequency synchronization accuracy over SIR_{in} for the detected communication signal with $\delta_w T_0 = \{-0.07, -0.15, -0.3\}$, consisting of the SCA outcome (sca), channel coefficient estimation (fine), to yield the total frequency offset (total).

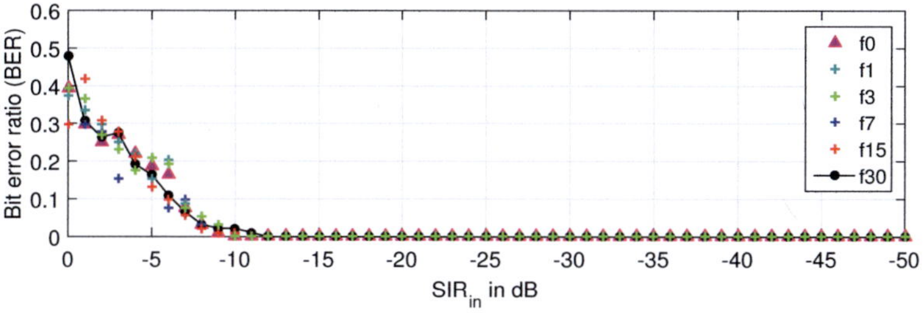

(c) Bit error ratio of the recovered Comm message signal.

Fig. 5.23: Case Study III: Simulated time and frequency synchronization accuracy over $\text{SIR}_{\text{in}} \in [-50, 0]$ dB as well as the BER of the recovered Comm message.

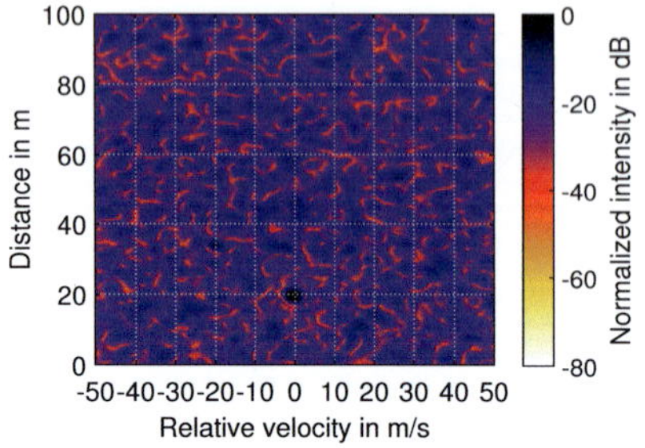

(a) Radar image with interference, $\mathrm{SNIR}_{\mathrm{out}} = 20.5\,\mathrm{dB}$.

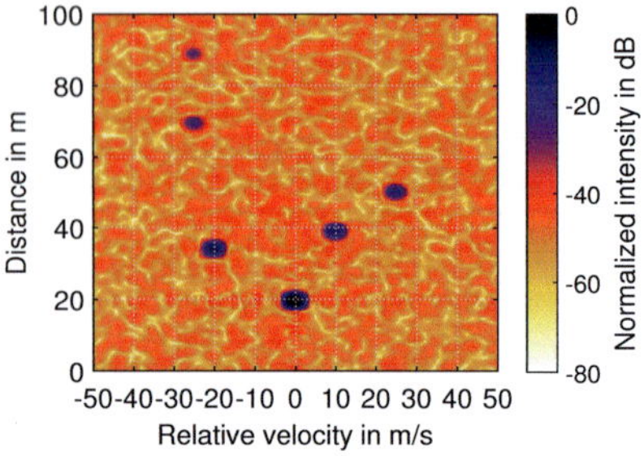

(b) Radar image after interference cancellation, $\mathrm{SNIR}_{\mathrm{out}} = 51.6\,\mathrm{dB}$.

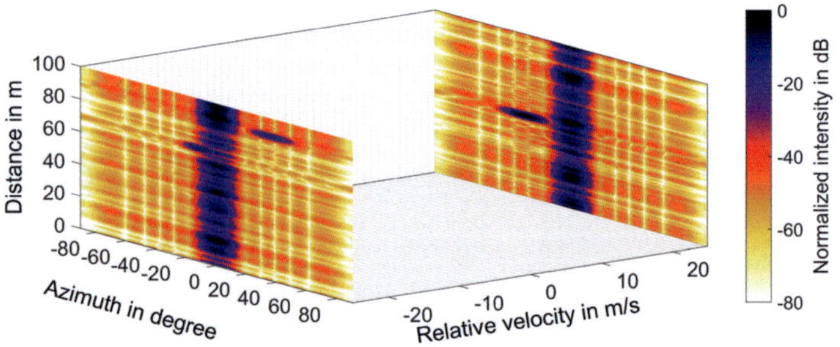

(c) DOA image (slice) with interference.

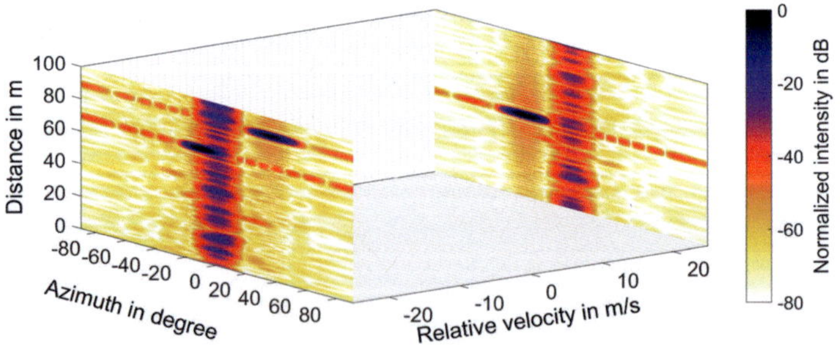

(d) DOA image (slice) after interference cancellation. T

Fig. 5.24: Case Study III: Simulated radar image and DOA image for $\mathrm{SIR}_{\mathrm{in}} = -30\,\mathrm{dB}$, $\delta_w T_0 = -0.16$ and $\mathrm{SNR} = 30\,\mathrm{dB}$ (same conditions as used for Fig. 4.8). Targets E and F are at -25 m/s and D at 25 m/s.

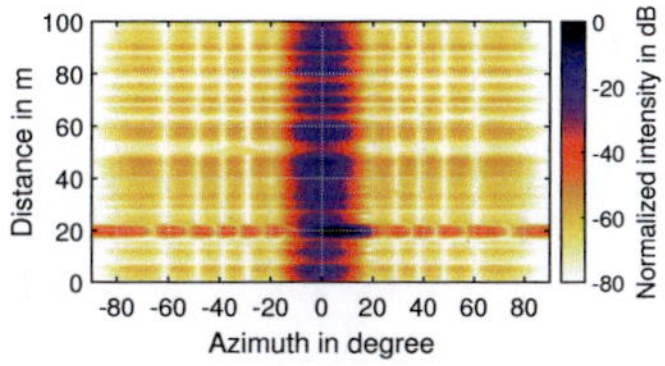

(a) A (0 m/s): interference.

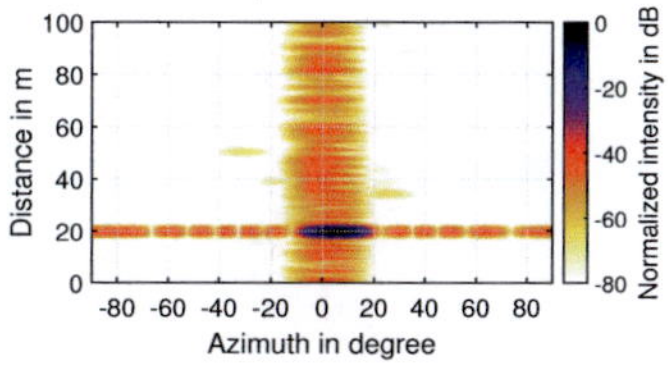

(b) A (0 m/s): canceled.

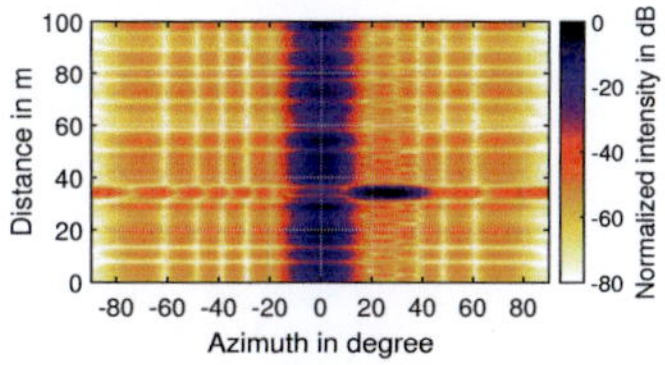

(c) B (−20 m/s): interference.

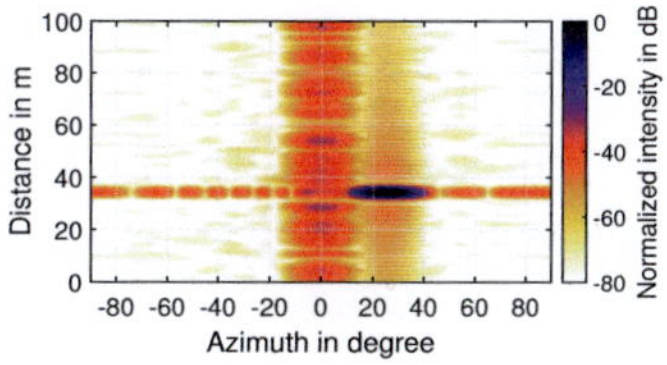

(d) B (−20 m/s): canceled.

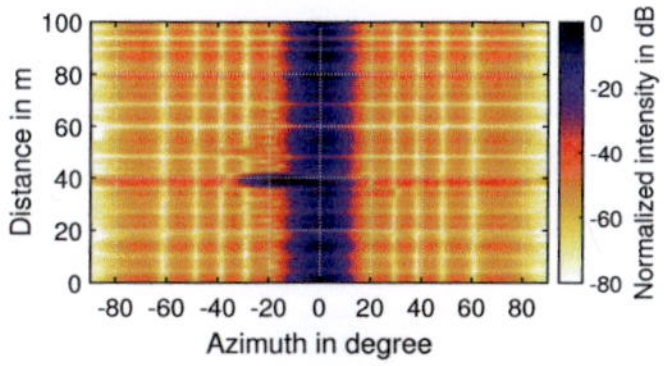

(e) C (10 m/s): interference.

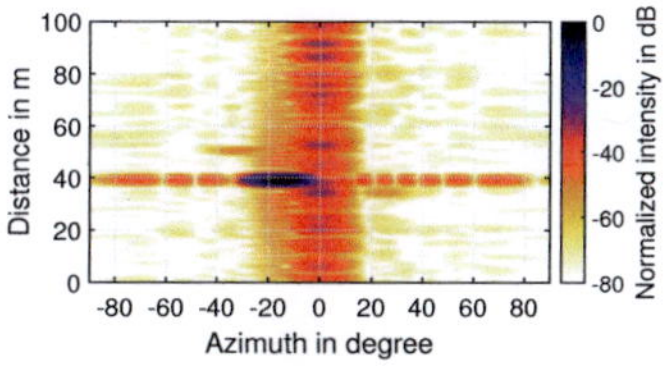

(f) C (10 m/s): canceled.

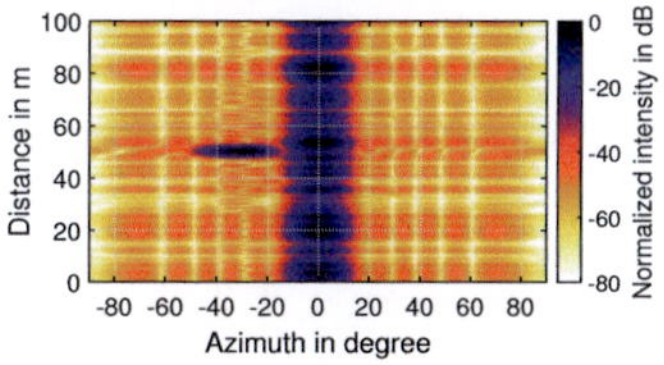

(g) D (25 m/s): interference.

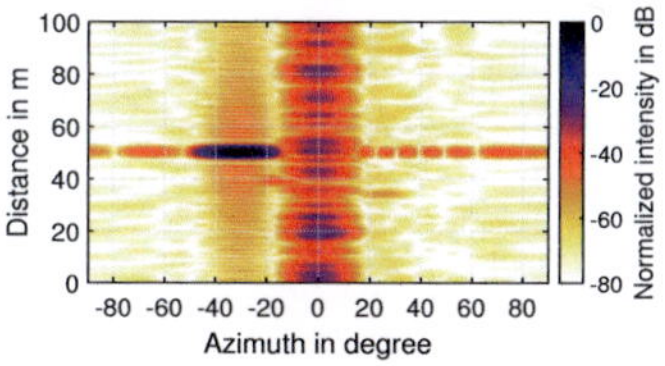

(h) D (25 m/s): canceled.

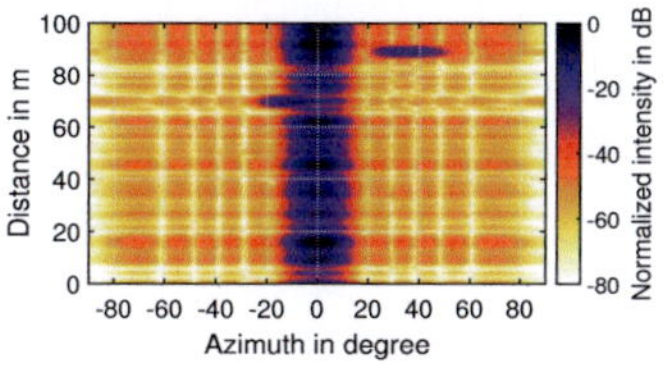

(i) E & F (−25 m/s): interference.

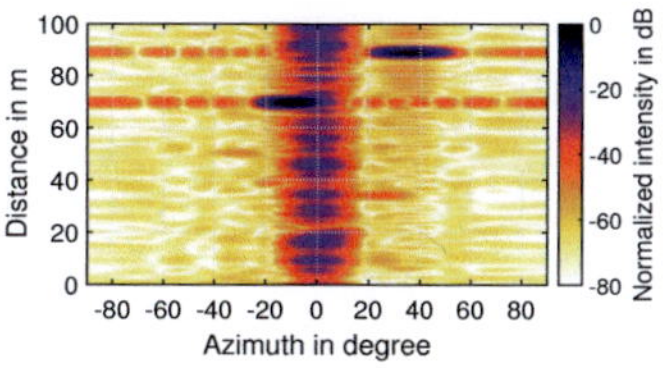

(j) E & F (−25 m/s): canceled.

Fig. 5.25: Case Study III: Comparison of simulated DOA image slices based on velocity according to the scenario given in Fig. 5.24.

5.9.1 Case Studies summary and conclusion

Table 5.3 provides a summary of the $SNIR_{out}$ of the Radar signal before and after interference cancellation according Case Study I, II and III for the same scenario depicted in all the radar images in Fig. 5.14, 5.19 and 5.24. In Case Study I, the subchannels of $u = 0$ that are farthest from the Comm's subchannels of $u = 4$ have the best isolation. The isolation decreases for the subchannels that are nearer and end with the worst isolation for the directly neighboring subchannels of $u = 3$. The interference canceled results also reflect this trend. At the best, the achievable $SNIR_{out}$ after interference removal is still 3 dB lower than the maximum achievable $SNIR_{out}$ in the scenario with no interference of 77 dB. Even if the time and frequency offsets can be synchronized to within the nearest correct time delay and frequency bins, the channel matrix remains somewhat distorted due to the presence of other signals leaking into the Comm's subchannels. Therefore it can be expected that the interference canceled result will never reach the maximum $SNIR_{out}$ of the scenario. As the subchannel misalignment grows, the worse the $SNIR_{out}$ after interference cancellation.

Table 5.3: Comparison of $SNIR_{out}$ between Case Study I, II and III for one Comm with $SIR_{in} = -30$ dB and $\delta_w T_0 = -0.16$.

user index,	Case Study I		Case Study II		Case Study III	
u	original in dB	canceled in dB	original in dB	canceled in dB	original in dB	canceled in dB
0	36	74	22	68	20	51
1	35	73	23	69	-	-
2	32	72	23	69	-	-
3	26	69	20	66	-	-

For Case Study II, where the Comm is a four-transmitter node, the subchannels of $u = 0$ are no longer the subchannels with the best isolation since the Comm's subchannels of $u = 7$ are directly adjacent to them. Hence the best isolated subchannels are those of $u = \{1,2\}$ but the isolation is only a mere 1 dB more. The subchannels of $u = 3$ bear the biggest brunt of the interference due to being directly adjacent to Comm's subchannels of $u = 4$ as well as the direction of the frequency offset that shifts the Comm's subchannels toward the direction of subchannels of $u = 3$. Due to the presence of multiple interfering Comm signals, the Radar's $SNIR_{out}$ is also much lower than in Case Study I as expected. Comparing the $SNIR_{out}$ of the subchannels of $u = 3$ for both Case

Study I and II, one can see that there is a 6 dB drop in $\mathrm{SNIR}_{\mathrm{out}}$, consistent with the Comm's four-times rise in power.

In Case Study III the $\mathrm{SNIR}_{\mathrm{out}}$ of the Radar's signal with interference from a one-transmitter Comm is a mere 20 dB, increasing to a $\mathrm{SNIR}_{\mathrm{out}} = 51$ dB, which is 17 dB worse than the four-transmitter Comm node case in Case Study II. *This implies that the classical OFDM signal model is not employable for a networked or multiple-user access RadCom application.*

Looking at the $\mathrm{SNIR}_{\mathrm{out}}$ figures of the Radar signal with interference and after interference cancellation, it is obvious from all the Case Studies that the spectrally interleaved OFDM signal model is clearly more robust than the classical model. Based on the $\mathrm{SNIR}_{\mathrm{out}}$ versus $\mathrm{SIR}_{\mathrm{in}}$ and frequency offset plots, the interleaved signal model can tolerate up to $\delta_w T_0 = 0.07$, or 7% of offset with respect to the subcarrier spacing Δf, whether with a single- or multiple-transmitter Comm node without much loss in $\mathrm{SNIR}_{\mathrm{out}}$ values after interference cancellation. In the Case Studies, this 7% off frequency offset is already equivalent to 6.36 kHz with $\Delta f = 90.9$ kHz.

The operation of the RadCom system, whether in a cooperative or non-cooperative network will require a certain knowledge about the state of the channels (i.e. channel availability), and hardware-wise they must already fulfill the requisite of frequency synchronization to allow only a tolerable margin of frequency errors. This concept is not new and is common to all OFDM-based communication systems including the new generation WiFi using 802.11g/n standards and LTE. A typical LO's accuracy is temperature dependent and may drift with an offset that can be as high as 10 ppm. Based on this LO imperfection and Doppler caused by moving user equipment, the offset from the carrier frequency must be corrected, usually at the downlink [WMR10].

Let us consider an example case. The LTE system requires the LO accuracy for a wide-area base station to be 0.1 ppm. If such an LO is available for use at 24 GHz, this will lead to an offset equivalent to 2.4 kHz. Now this offset value has yet to take RadCom node and/or target movements or velocities into account. If employed in an automotive scenario, where the maximum speed can be assumed to be 200 km/h (55.6 m/s), the maximum frequency offset due to Doppler is then 4.5 kHz at $\Delta f = 90.9$ kHz. Taking into account the LO offset and the Doppler, this sets the total frequency offset at 6.9 kHz (or 7.6% of the subcarrier spacing $\Delta f = 90.9$ kHz), still somewhat within the aforementioned tolerance region after interference cancellation. Hence this shows that the expected frequency offset value in a real scenario is in the kilohertz range.

It can thus be surmised that the interference cancellation algorithm will definitely be required for the reliable operation of the RadCom in a network or multiple-user scenario. However, in order for it to be operable in a real-time scenario, a low-complexity yet efficient algorithm must be used. The interference cancellation algorithm presented in this thesis can be adapted to run in parallel for all Rx. Currently with a regular personal computer (Intel i5 CPU with 3.33 GHz frequency 16 GB RAM) with no code optimization in Matlab, the interference cancellation algorithm takes about 10 s per Comm signal. With a server-type processor (i.e. Intel Xeon) this time is reduced to 4 s per Comm signal.

5.10 Chapter 5 summary

All RadCom nodes use different SCA preambles, but the preambles are the same for all frames of the same node. The first stage time synchronization detects pairs of peaks to confirm the start point of the Comm signal. Under certain circumstances, this can give a false result, hence the second detection stage is necessary. The SCA estimated the frequency offset at the frequency-subcarrier axis of the frame-under-evaluation. When there are multiple overlapping signals from multiple Comms or multipath, the SCA tend to give the estimate of the vector sum (of positive or negative values) of the frequency offset. Again a second stage frequency estimation and synchronization becomes necessary. If the SCA can correct most of the frequency offset (a coarse correction), the second stage will be able to detect the fine residue frequency offsets and this lead to a cleaner channel equalization matrix as well.

In real cases where there are multiple Comms and the physical orientation of the Comms toward the Radars are taken into account, the interference cancellation algorithm must be done on every of the Radar's Rx to detect each Comm or channel for cancellation. This is because the path length from every Comm's transmit antenna to every Radar's receive antenna is different and this affects the channel equalization matrix. To enable real-time processing of the interference cancellation algorithm, the proposed algorithm is sufficiently low in implementation complexity and can be optimized for a fast processing.

In the Case Studies presented, it is obvious that the spectrally interleaved signal model is much more robust for a multi-antenna, multi-user scenario. It has been demonstrated that the classical OFDM signal model is not able to achieve a significant $SNIR_{out}$ and in the case of more than one Comm node or a Comm node with multiple antennas, this signal model will not be able to achieve a sufficient $SNIR_{out}$ for a reliable radar estimation even with interference cancellation. As such, the classical OFDM signal model is not employable in real scenarios whereas the spectrally interleaved OFDM signal model is a prime candidates for use in MIMO Radar systems.

6 Application level: Subcarrier allocation and target tracking

At this point, adequate insights have been presented about the characteristics of the chosen spectrally interleaved OFDM signal model, the physical antenna requirements, and the tolerance of the MIMO RadCom toward mutual interference. It is already known that assigning non-overlapping channels to all RadCom nodes in a network is the first line of mutual interference mitigation. However it has also been seen that the user index imposes an extra phase offset term in Section 2.4.2, which is constant (affecting every modulation symbol by the same amount) over every qp-th OFDM frame. This inadvertently also influences the accuracy of the DOA estimation to a certain extent. To complete the picture of the system, this chapter will discuss the subcarrier allocation strategy and show its effect on the DOA image.

The second part of this chapter then presents a simulation of a complete system radar toy model that performs target tracking using a simple Kalman filter. This is intended to give an insight to the possibilities of the MIMO RadCom especially in the area surveillance applications.

6.1 Subcarrier allocation among multi-antenna radar nodes

Up to this point of the thesis, whenever DOA images are presented, the subcarrier assignment used to generate the images were always 'regular' in the sense that the user index assigned to the Radar were always contiguous, dictated by $u = 0, 1, 2, 3$ for a 4×4 configuration. The regularity of the subcarrier assignment has proven to affect the radar estimation the least, as compared to a 'random' subcarrier assignment. In [Bra14], a random subcarrier allocation refers to a fair but random selection of subcarriers to be assigned to any one RadCom node. Radar estimation from a non-regularly sampled data causes the radar periodogram to suffer from random spurs. Besides, extra synchronization must be put in place at the Rx to detect the subcarriers used by the other RadCom nodes[16], hence the disadvantage of this method clearly outweighs the advantage.

In order to see the impact of the signal model on the DOA estimation, the Fourier Beamforming equation (Eq. (3.41)) will be revisited. Assuming that the received signal has been normalized by the correct beamsteering vector and contains no other phase offsets from hardware ($\mathrm{e}^{\mathrm{j}\zeta} = 1$), only the relevant terms from Eq. (2.26) (the modulation symbols at the q-th Rx) with the DOA term as given in Eq. (2.37) will remain. The Fourier Beamforming DOA results from Eq. (3.41) can now be expanded as

$$\begin{aligned}\mathbf{FB}(n,m) &= \sum_{h=0}^{H-1} \mathrm{e}^{\mathrm{j}\Psi_{p,q,h}} \mathrm{e}^{\mathrm{j}\varphi_{\mathrm{R}_h}(u)} \mathrm{e}^{\mathrm{j}\varphi_{\mathrm{train}}} \cdot \mathrm{IDFT}\left[\mathrm{DFT}\left[\vec{\kappa}_{r_h} \otimes \vec{\kappa}_{\mathrm{D}_h}\right]\right] \\ &= \sum_{h=0}^{H-1} \mathrm{e}^{-\mathrm{j}2\pi\frac{|r_{p,h}+r_{q,h}|}{\lambda_c}} \cdot \mathrm{e}^{-\mathrm{j}2\pi u \Delta f \tau_{p,q,h}} \cdot \mathrm{e}^{\mathrm{j}\varphi_{\mathrm{train}}} \cdot \mathrm{IDFT}\left[\mathrm{DFT}\left[\vec{\kappa}_{r_h} \otimes \vec{\kappa}_{\mathrm{D}_h}\right]\right].\end{aligned} \tag{6.1}$$

As can be seen, the first two exponential terms cast a fixed phase offset that affects every modulation symbol in the qp-th signal frame and cannot be estimated or equalized through the Fourier-based range-Doppler estimation and beamsteering vector (while $\mathrm{e}^{\mathrm{j}\varphi_{\mathrm{train}}}$ is the same over all signal frames hence it has no influence). The first term is the desired DOA term whereas the second one is the term caused by the user index. These terms will add up to cause a slight shifting of the angle estimation. The parameters that govern the amount of shift in the angle estimation are:

- the physical antenna arrangement i.e. the contiguity of the array elements
- user index u of the signal assigned to the above-said antennas
- the distance of the object given by its corresponding time delay $\tau_{p,q,h} = \frac{r_{p,h}+r_{q,h}}{c_0}$

[16] carrier sensing to detect subcarriers used by nearby RadCom nodes

To simulate the effect, a scenario with one target located at $\phi = 0°$ with an RCS of $10\,\mathrm{m}^2$ is set up in Matlab. The Radar is an 8×4 node and the signal bandwidth is divided into $N_{\mathrm{ch}} = 8$ channels (making the maximum unambiguous range 206 m). The target is simulated at a few ranges as given by $r = \{10, 50, 100, 150, 200\}$ m, up to the radar's maximum unambiguous range. At any one time, the Radar will only use four Tx, which can be chosen among the ULA transmit antenna indices of $p \in \{0, 1, 2, 3, 4, 5, 6, 7\}$. The four Rx indices are fixed at $q \in \{0, 1, 2, 3\}$ and are centered around the origin (middle point of the array) with $d_{\mathrm{R}} = \lambda_{\mathrm{c}}/2$ while $d_{\mathrm{T}} = 2\lambda_{\mathrm{c}}$. The other simulation parameters as as shown in Table 5.2.

To show the effects of the aforementioned parameters, the simulation is divided into two antenna configurations. The first one uses contiguous antenna elements, while the second one uses every other Tx element (from the total available eight antennas). Within these antenna configurations, the effect of the user index allocation will be shown.

6.1.1 Contiguous transmit antennas

The locations of the transmit and receive antenna elements and the resulting virtual array are as shown in Fig. 6.1. When the transmit antenna elements of the contiguous indices $p = \{2, 3, 4, 5\}$ are used, a contiguous virtual array also results. With such a configuration, the azimuth cuts of the DOA images (with Hamming windowing) without the influence of the user index are overlaid together is as shown in Fig. 6.2(a). The plot shows one clear peak with no grating lobes in the $\phi = [-90°, 90°]$ region. This is considered an optimal DOA estimation result.

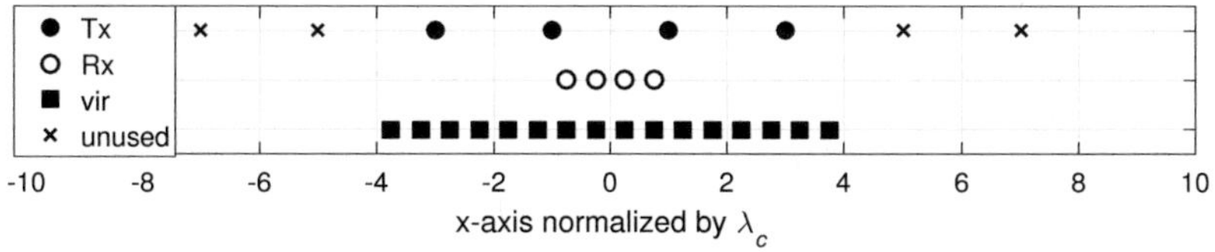

Fig. 6.1: Contiguous: Position of the Tx, Rx and virtual array ('vir') elements.

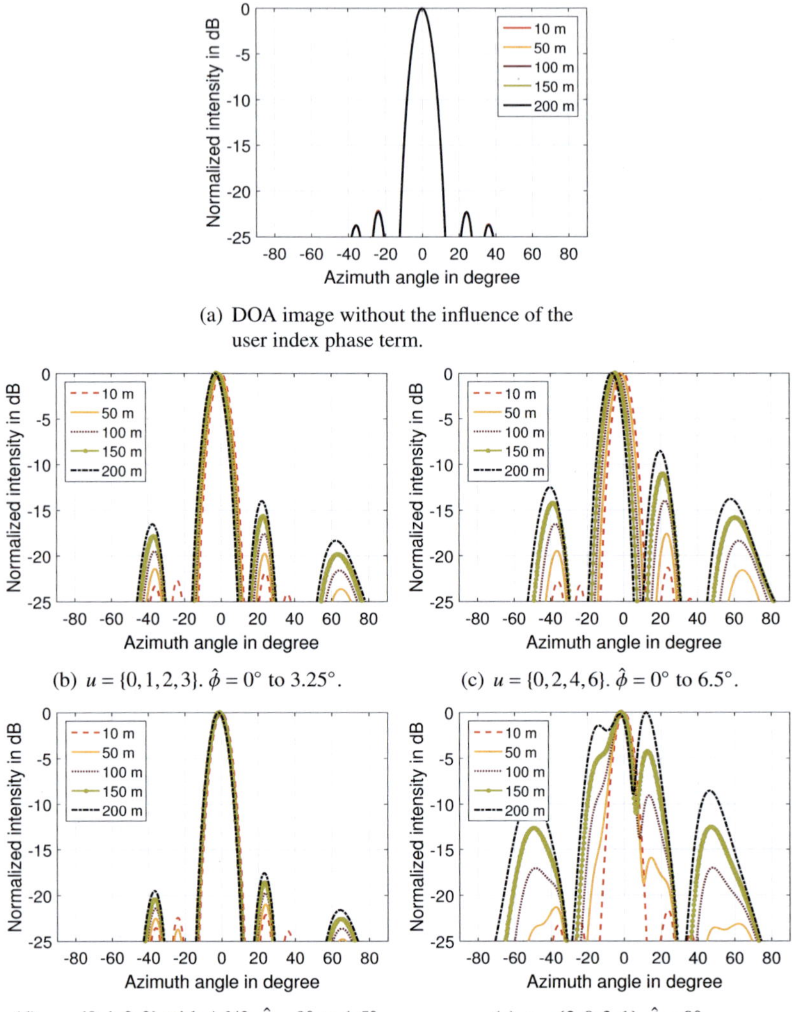

(a) DOA image without the influence of the user index phase term.

(b) $u = \{0, 1, 2, 3\}$. $\hat{\phi} = 0°$ to $3.25°$.

(c) $u = \{0, 2, 4, 6\}$. $\hat{\phi} = 0°$ to $6.5°$.

(d) $u = \{0, 1, 2, 3\}$ with $\Delta f/2$. $\hat{\phi} = 0°$ to $1.5°$.

(e) $u = \{2, 0, 3, 1\}$. $\hat{\phi} \approx 0°$.

Fig. 6.2: Contiguous: Resulting DOA estimation (with Hamming window) for a 4×4 1D collocated transmit $p = \{2, 3, 4, 5\}$ and received antennas for one target at $r = \{10, 50, 100, 150, 200\}$ m and $\phi = 0°$. $\hat{\phi}$ is the estimated peak location.

Next, four different cases of the user index assignment to the transmit antennas will be analyzed. In the first case of allocating $u = \{0,1,2,3\}$ to $p = \{2,3,4,5\}$ respectively or in a reverse order, Fig. 6.2(b) results. At 10 m the DOA angle estimate is 0° but as the distance becomes larger, the effect of $e^{j\varphi_{R_h}(u)}$ also becomes more prominent and as a consequence, the target detection is also inflated by $\hat{\phi} = 3.25°$ at $r = 200$ m. Since this angle deviation is still within the angular resolution, it is not considered as a minus point to the RadCom. It can also been seen that the amplitude of the sidelobes increases as the range grows larger. If the u is allocated in the reverse order to the Tx antennas, the angle deviation of the target in Fig. 6.2(b) will grow toward the right side instead.

In Fig. 6.2(c), $u = \{0,2,4,6\}$ is assigned to $p = \{2,3,4,5\}$ respectively. This in essence adds a factor of 2 to the $e^{j\varphi_{R_h}(u)}$ hence the estimated angle deviation is also twice as large as from the aforementioned first case, with the angle deviation $\hat{\phi} = 6.5°$ at $r = 200$ m. Since $e^{j\varphi_{R_h}(u)}$ is also dependent on the subcarrier spacing, using $u = \{0,1,2,3\}$ while halving Δf will also improve the angle deviation by half (at the expense of halving the signal bandwidth) as shown in Fig. 6.2(d) as compared to Fig. 6.2(b), where largest deviation is $\hat{\phi} = 1.5°$.

The final case is a random user index allocation of $u = \{2,0,3,1\}$ as shown in Fig. 6.2(e). In this case the angle estimate is almost the same $\hat{\phi} \approx 0°$, even over large distances but at the expense of the erratic sidelobes. The larger the distance, the larger the sidelobes. This allocation is not suitable for use for radar estimation over 10 m since the immediate sidelobes are too high despite the Hamming windowing and will potentially cause false positive targets.

While at the first look at the results it may seem that the spectrally interleaved OFDM signal model suffers from inaccurate DOA estimations, a few real world parameters must be taken into consideration as follows, which would show the boundaries of its utilization.

- Although it is not obvious from Eq. (6.1), the higher the number of transmit and receive antennas used, the lower the angle deviation at the farthest range will be. For example, with an 8×8 configuration, the DOA angle deviation at $r = 200$ m decreased to 2° instead of 3.5°.
- In a real scenario, a target is not a point scatterer as used in the simulation here. A common target used for radar calibration, the trihedral corner reflector also exhibits various points of reflections[17] depending on the direction of the radar's incident waves toward its axis of symmetry.

[17] the mechanism in explained in [Rob47]

Hence it is common for a target to show up as a clutter of points on the radar and DOA (with Fourier Beamforming) images instead of a single point. In this respect, a deviation of a mere 3.5° will not affect the DOA estimation at all since the clutter of points will spread over a ±10° area of the azimuth axis. This will be demonstrated in the measurements in Chapter 7.

- The term $\mathrm{e}^{-\mathrm{j}2\pi u\Delta f\tau_{p,q,h}}$ in Eq. (6.1) that causes the angle deviation is dependent on the symbol duration T_0 or subcarrier spacing Δf of the OFDM signal. These are also related to the maximum unambiguous range $r_{\max}$ of the radar, whereby a longer T_0 is proportional to a larger $r_{\max}$. To enable a correct DOA estimation without ambiguities, the range of the farthest target must fall within $r_{\max}$. Therefore for a long range target estimation for example, the symbol duration must be lengthened to $6 \cdot T_0$ to result in $r_{\max} = 1238$ m. In this case the angle deviation at $r_{\max}$ is a mere 1.5° for an 8×8 configuration and 3° for a 4×4 configuration. Increasing T_0 by six times however has the consequence of lowering the range resolution Δr to a directly proportional value. To retain the Δr, the signal bandwidth must be also increased by six times.
- A post-processing phase compensation can be done, where each OFDM processed matrix is multiplied with the conjugate of the $\mathrm{e}^{-\mathrm{j}2\pi u\Delta f\tau_{p,q,h}}$ term for all targets.

Looking at the arguments above, it is clear that the question of the DOA inaccuracy can be circumvented with a smart parameterization of the spectrally interleaved OFDM signal according to the application and hardware limits or by employing a post-processing algorithm. When used in a radar network for area surveillance for example, a time-division multiplexed scheme can be employed so that not every RadCom node will be transmitting at the same time, thus lowering number of channels N_{ch} required and hence extending the $r_{\max}$ for all nodes.

6.1.2 Sparse transmit antennas

In the same manner, the antenna configuration with $p = \{0,2,4,6\}$ will be analyzed. This is akin to having transmit antennas that are spaced twice the element spacing away from each other compared to the contiguous Tx antenna case. Fig. 6.3 shows the locations of the transmit antennas and the resulting sparse virtual array. The DOA azimuth cut in Fig. 6.4(a) features a sharp main lobe with very high sidelobes despite the use of a Hamming window. If this configuration were to be used, the unambiguous angular FOV will be confined to around $\phi = [-10°, 10°]$ only, narrowly avoiding the high sidelobes at $\hat{\phi} = 14.5°$. Suffice to say, this is not an employable configuration for multi-target estimation. Nevertheless, the effects of the subcarrier allocation will be presented for comparison.

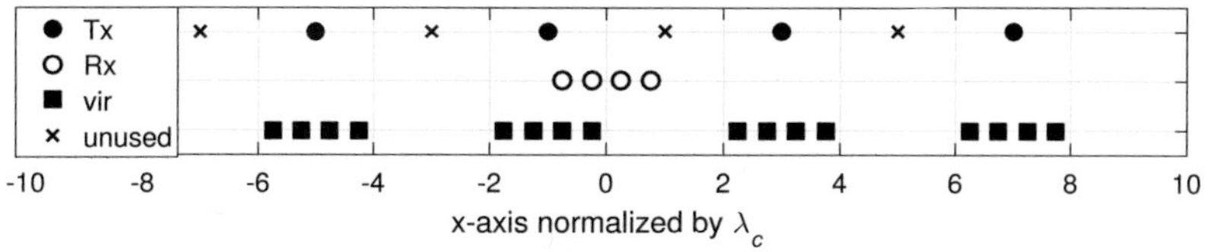

Fig. 6.3: Sparse: Position of the Tx, Rx and virtual array ('vir') elements.

Fig. 6.4(b) shows the result for the $u = \{0, 1, 2, 3\}$ assignment to the transmit antennas $p = \{0, 2, 4, 6\}$. As expected of the narrower main lobe, the angle estimate deviation is also more than a factor of two smaller than for the contiguous antenna case, at $\hat{\phi} = 1.75°$. The same can be said for Fig. 6.4(d) also using the $u = \{0, 1, 2, 3\}$ allocation but with half the subcarrier spacing, where the estimated angle deviation is $\hat{\phi} = 0.75°$. While both of these configurations feature excellent DOA angle estimates, the high sidelobes of -4 dB are a clear disadvantage, which severely limits the DOA unambiguity region.

When using the $u = \{0, 2, 4, 6\}$ subcarrier assignment, the angle estimate deviation doubles to $\hat{\phi} = 3.5°$, which is as expected, as shown in Fig. 6.4(c). Finally with the random subcarrier allocation of $u = \{2, 0, 3, 1\}$, the DOA image in Fig. 6.4(e) is rendered very ambiguous with multiple peaks resulting for one sole target. Besides the very limited angular FOV, the DOA estimation of a target at all distances will interfere with the angular resolution. It can hence be concluded that this transmit antenna element and subcarrier allocation strategy is not employable in any case.

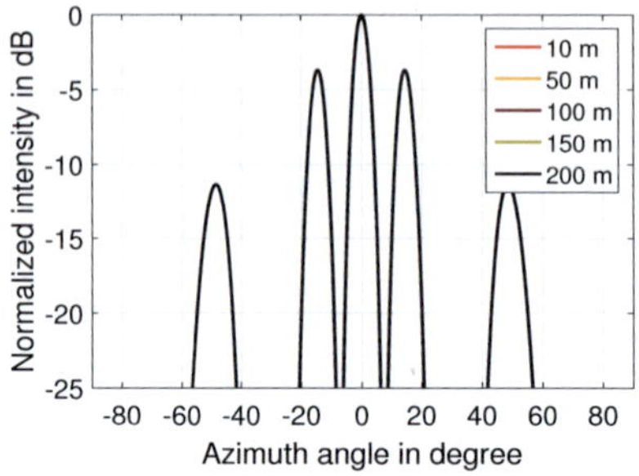

(a) DOA image without influence of the user index phase term.

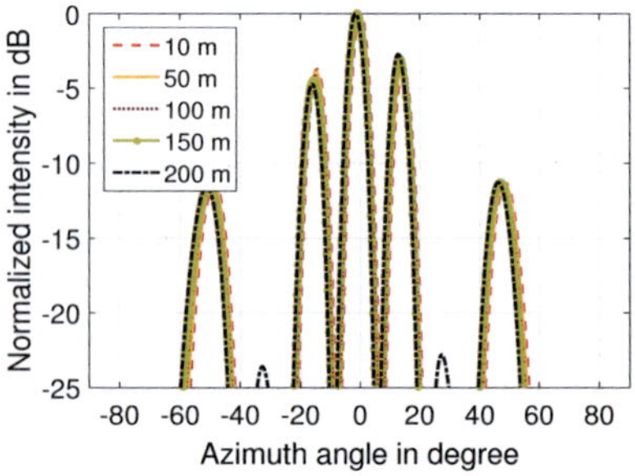

(b) $u = \{0, 1, 2, 3\}$. $\hat{\phi} = 0°$ to $1.75°$.

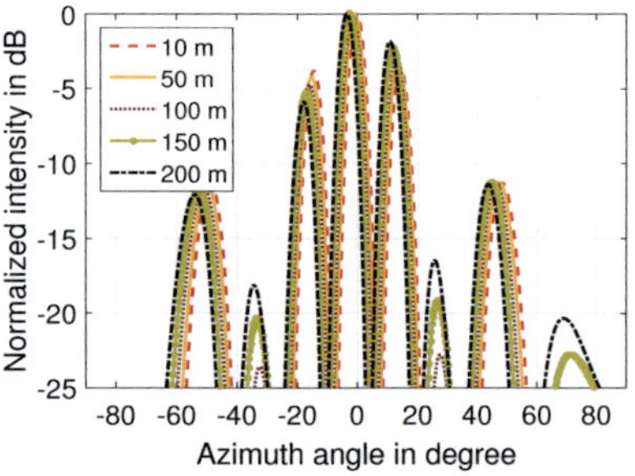

(c) $u = \{0, 2, 4, 6\}$. $\hat{\phi} = 0°$ to $3.5°$.

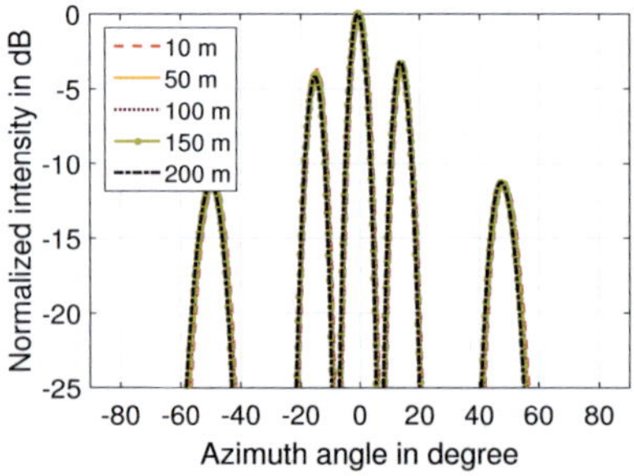

(d) $u = \{0, 1, 2, 3\}$ with $\Delta f/2$. $\hat{\phi} = 0°$ to $0.75°$.

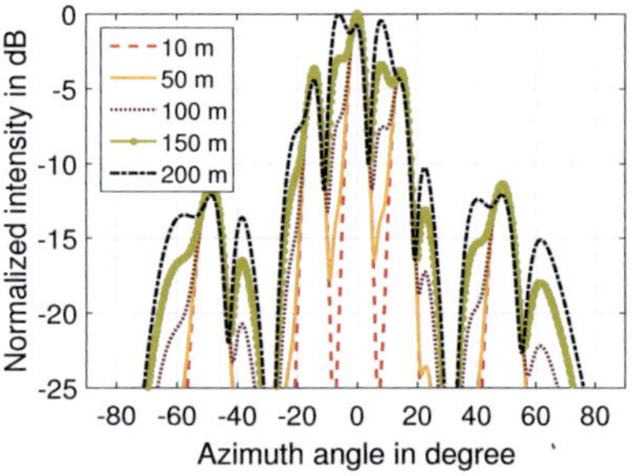

(e) $u = \{2, 0, 3, 1\}$. $\hat{\phi} \approx 0°$.

Fig. 6.4: Sparse: Resulting DOA estimation (with Hamming window) for a 4×4 1D collocated transmit $p = \{0, 2, 4, 6\}$ and received antennas for one target at $r = \{10, 50, 100, 150, 200\}$ m and $\phi = 0°$. $\hat{\phi}$ is the estimated peak location.

6.2 Radar Toy Model for target tracking

Following the discussion thus far, it is evident that with the correct subcarrier allocation strategy and physical antenna geometry the MIMO RadCom is able to estimate the 3D+velocity parameters within one transmit cycle that is usually within the span of milliseconds. In terms of functioning as an area surveillance radar however, one big question remains, i.e. how does the radar distinguish an arbitrary number of targets that it is observing? Consider in real scenarios that the targets will also be moving, with movements that can either be predictable or erratic. Besides, there will be clutter from certain directions that will show up in the radar estimation as ghost targets. One of the simplest ways to distinguish multiple targets is to track them over a span of time and associate their trajectories based on their speed.

Tracking the targets would mean transmitting signals periodically to gauge the change in the targets' locations. After a minimum of two radar estimations are done, the changes that are recorded can be used as a prediction of target velocity and direction. Then estimating the number of targets present in the area of surveillance and differentiating them by their trajectories over time will allow them to be distinctly distinguished.

It is the intention here to show the simulation of a simple tracking algorithm (published in [SSWZ15]), which can be used with the proposed MIMO RadCom of this thesis. For that reason, a radar toy model is set up for simulation at 24 GHz with the parameters as shown in Table 5.2. The parameters are intended for automotive radars, which can be implemented in real time with Software Defined Radios (SDR) but they can be easily modified for other applications. The SDR envisioned for use is the latest ETTUS X310 universal software radio peripheral (USRP), which can achieve a 100 MHz instantaneous analog bandwidth albeit at a lower operating frequency. The toy model comprises 8×8 1D (azimuth only) ULA antennas with $d_{\mathrm{R}} = \lambda_{\mathrm{c}}/2$ and $d_{\mathrm{T}} = 4\lambda_{\mathrm{c}}$. This will result in a contiguous 64-element 1D virtual array with no over-representation of the same receive signal (no redundancy) as explained in Chapter 3. The total transmit signal duration is $12.375\,\mu\mathrm{s}\cdot M = 3.168$ ms. Assuming that the furthest object is located at the radar's maximum distance of 206 m (the maximum unambiguous distance possible when $N_{\mathrm{ch}} = 8$), only 1.37 µs is required for the reflected signal to arrive in a round-trip.

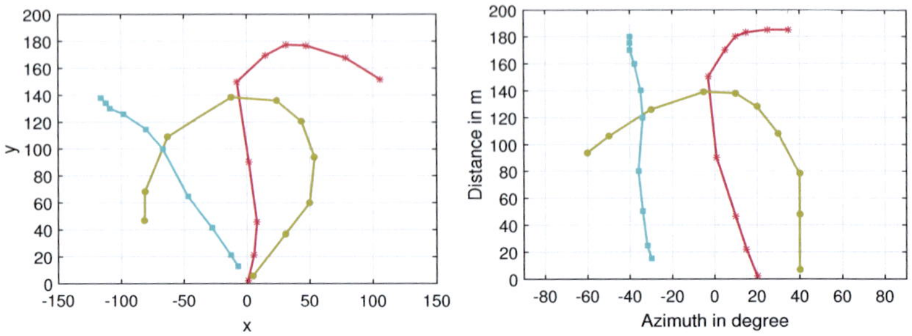

(a) True trajectory in Cartesian coordinates.

(b) True trajectory in polar coordinates.

Fig. 6.5: True trajectories of the three targets indicated by the different colors moving at different velocities. Each point represents the location of the respective target after the scan time of 2 s.

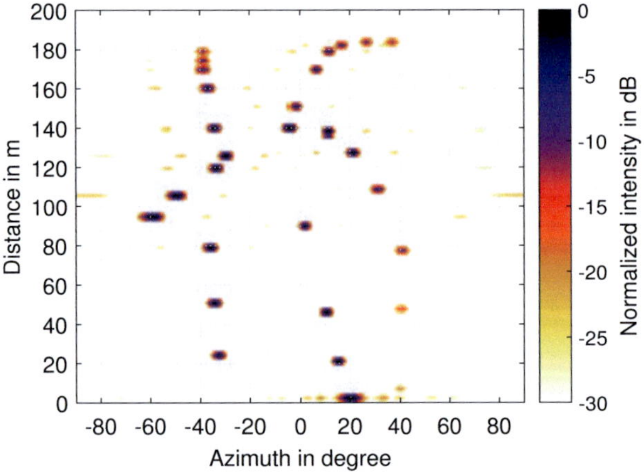

Fig. 6.6: DOA pseudo spectrum for three targets with different velocities in 10 snapshots of over a period of 18 s.

Taking into account the transmit signal duration, the time required for the USRP in between snapshots (system group delay), and the expected largest velocity of the target under observation, the scan time (or time interval between radar snapshots) is chosen to be 2 s. Ideally the radar processing is to be done in parallel while the next transmit signal is sent. The simulation is executed serially in Matlab on a regular personal computer and takes around 0.8 s per snapshot with no zero-padding during the range and velocity estimation. Hence the resulting DOA image is of low accuracy[18].

A typical scenario is set up to consist of three point-scatterer targets of the same RCS. All three targets start out at slightly different distances from the radar, from 2 to 10 m away, with velocities (relative to the Radar) ranging from 0 to ±30 m/s. These targets emulate the speed and trajectory of cars as shown in Fig. 6.5. The Radar's output is in the spherical coordinates while the tracking algorithm uses the Cartesian coordinates.

A total of 10 radar snapshots from the time 0 s to 18 s are taken. After each snapshot, the processed $\mathbf{I}(k,l,\phi)$ matrix is passed to the tracking system. The superimposed processed results from all $\mathbf{I}(k,l,\phi)$ matrices over time are as shown in Fig. 6.6. The lighter intensity trails that accompany the maxima are the unavoidable sidelobes that result from the Fourier Beamforming step even with a Hamming window.

6.2.1 Target tracking

After processing each snapshot to obtain the DOA image, tracking is done to extract the target trajectories and improve the positioning accuracy. There are a few processes that should be done before performing the tracking. Fig. 6.7 shows the overall view of the system functions at each scan time.

The aim of the detection stage is to decide whether any target is present in the area based on the DOA estimations. For this purpose, a threshold is first applied on every DOA image to extract the potential target positions. This step usually generates more than one position per target. Clustering is then performed to assign close points with each other. Here, all points are clustered into groups of points based on their relative distances. A point falls into a cluster if its distance from the mean of the points currently assigned to that cluster is lower than a predefined value. The average position of each cluster is considered as the final estimated position, called the *observation*.

A threshold of 0.2 of the maximum signal strength (maximum intensity of the radar output is normalized to 0 dB) is chosen here, whereby all signals with a power of higher than the threshold will be detected. It must be pointed out here that the tracking is performed in the Cartesian coordinates hence the estimates

[18] since without zero-padding, a sharp peak cannot be yielded

from the DOA image will be converted to the Cartesian coordinates accordingly. The clustering distance is defined as 10 m. This results in the detected points and clusters shown in Fig. 6.8(a) in Cartesian. The polar coordinate equivalent is shown in Fig. 6.8(b). The detected points ('detect') that are above the threshold form a cluster. The observations ('observ') are shown as a big dot in the center of the cluster of points. It can be seen that these observations correspond well to the true trajectories ('true' in solid lines).

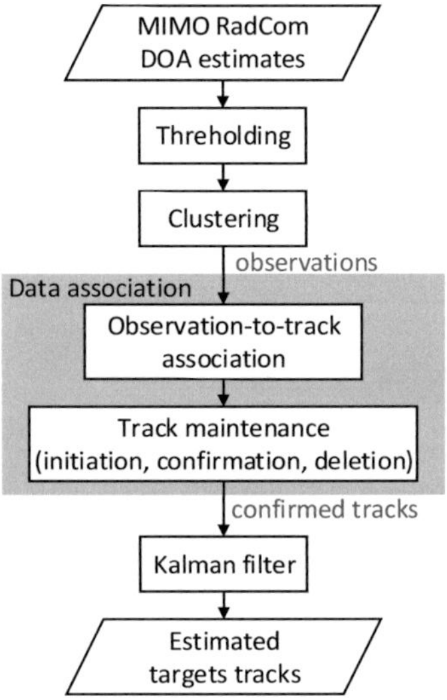

Fig. 6.7: Tracking system flow.

The outputs of the clustering are then passed to the data association step, which associates the observations to the corresponding tracks. It is responsible for initiating tracks, forming potential tracks, confirming tracks and deleting them. For a real-time implementation feasibility, a simple and most commonly used data association logic in [BP99] is adopted to avoid further algorithm complexity. In this technique, all potential tracks are constructed at each scan time and under certain conditions they are converted into confirmed tracks. A *potential* track is a sequence of observations over time that are likely to be a track, whereas a *confirmed* track is certainly a track.

An observation is assigned to a (potential or confirmed) track if it falls inside a gate around the previous observation assigned to that track. The radius of the gate is defined according to the target's maximum velocity. If a potential track reaches a certain number of assignments, it is then converted to a confirmed track. If a (potential or confirmed) track is not assigned any observation (because it is missed) for a certain number of times, it is deleted.

Considering the above logic, three cases may happen for a given observation. If the observation falls into the gate of only one track, it is then assigned to that track. If the observation falls into the gate of more than one track, it is assigned to the one that is nearest. If the observation does not fall inside the gate of any track, then it starts a new track. Here the gate radius is set as 75 m, according to the maximum expected change of distance possible within the scan time of 2 s. If a track is missed for three times, it is deleted. A potential track is confirmed when it is assigned at least three consecutive non-missed observations. The results are as depicted in Fig. 6.8(c) and Fig. 6.8(d), whereby the tracks ('assoc' with dashed lines) for each of the three targets are connected through their observations.

At each scan time, the confirmed tracks are passed to the tracking filter, which recursively predicts the targets' positions and corrects the predicted positions based on the current observations. Here, the well-known Kalman tracking filter, which assumes a linear/Gaussian system model, is used. More efficient tracking approaches such as nonlinear Bayesian filters described in [AMGC02] can also be used at the cost of a higher computational complexity. However, the Kalman filter allows a fast processing for real time applications by performing linear operations, while providing a satisfactory tracking performance. For the Kalman filter, the acceleration noise and measurement noise standard deviations are set to 7 m/s^2 and 10 m (this value is the minimum change in distance of the slowest target) respectively. The results are shown in Fig. 6.8(e) and Fig. 6.8(f). The average processing time required to perform the tasks described in Fig. 6.7 is approximately 0.01 s per scan.

It can be seen that the tracked results ('track' in solid lines of different color) in Fig. 6.8 match well with the true trajectories albeit with an slight error at two of the start points due to the weak target intensities of below the threshold value at the detection stage. This is commonly called the shadowing effect, where the objects further behind a large RCS target appear as weak targets to the radar. From Fig. 6.6, it is evident that the targets and their tracks cannot be identified simply by looking at a time-series of snapshots hence the importance of the post-processing tracking system. Although clutter and non-target objects were not included in this simulation, they will require just an additional step using the Moving Target Indicator method to subtract the static clutter or non-targets from the radar image before the tracking step.

To detect human beings or objects within a shorter range with better position accuracy, a wider bandwidth can be used for a better range resolution. The transmit signal duration or the frequency of operation can be made longer or higher respectively for a finer Doppler resolution. More antennas can also be used for better separability of targets in the azimuth, else super-resolution algorithms such as MUSIC [Sch86] can be used to aid the target detection and tracking. All these methods will come at the price of extra complexity and time.

The MIMO RadComs can also be strategically placed to surround the area of observation and be networked together to minimize the shadowing effects. A 3D data fusion algorithm using the 3D+velocity estimation outcome of the networked MIMO RadCom nodes has been presented in [NSZ16].

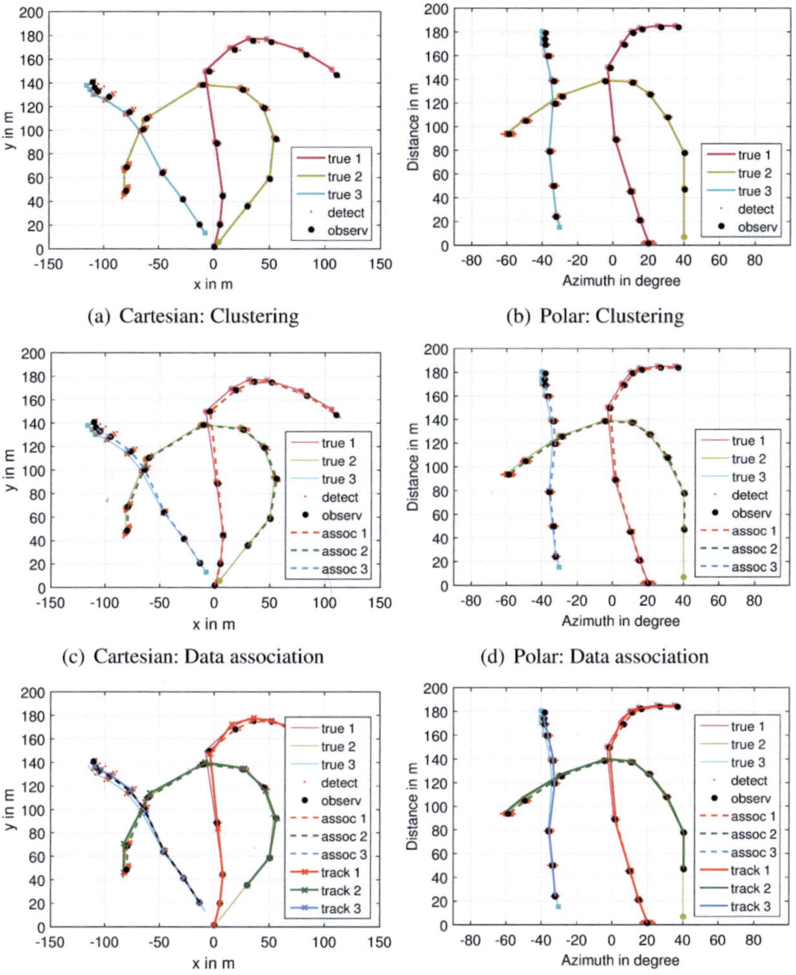

(a) Cartesian: Clustering

(b) Polar: Clustering

(c) Cartesian: Data association

(d) Polar: Data association

(e) Cartesian: Kalman filter tracking

(f) Polar: Kalman filter tracking

Fig. 6.8: Tracking system for three targets with different velocities in 10 snapshots over a period of 20 s, in Cartesian and polar coordinates. 'True' refers to the true trajectories of the targets from Fig. 6.5. 'detect' are the points of the DOA estimate above the set threshold value. 'observ' is the observation (average location of the points). 'assoc' refers to the data association track while 'track' is the result of the Kalman filter.

6.3 Chapter 6 summary

For an optimal performance of the MIMO RadCom using the spectrally interleaved OFDM signal model along with antenna array, the user index assigned to the antennas and the antenna elements used must both be contiguous. This is especially relevant in cases where there are more antennas than channels hence some antennas need to be disabled as per the example in Section 6.1. Smart parameterization of the RadCom system i.e. the number of antennas, bandwidth, symbol duration, etc. can be applied to further reduce the angular estimation error due to influence of the user index.

An example application utilizing post processing methods with the proposed MIMO RadCom has been shown. When multiple nodes are combined, the estimation is foreseen to have higher target localization accuracy. Arranging the nodes to surround the area of observation will reveal targets that are concealed from the view of some RadCom nodes due to terrain or obstacles through the data fusion from the receive signals of the nodes.

7 System setup and verification measurements

The verification measurements for the proof of concept of the 3D+velocity estimation capability of the radar as well as the interference cancellation algorithm are performed in real-time using the ETTUS X310 USRPs with 120 MHz daughterboards. For the first time in literature, a real-time 100 MHz bandwidth, 4×4 MIMO RadCom using parallel transmission with commercially available hardware is reported. While many researches concerning MIMO Radars are published on an ever increasing scale, to date, only Pfeffer et al. in [PFS15] in late 2015 mentioned a working MIMO Radar hardware called the Software Defined Radar System (SDRS). The SDRS is custom-built with industrial cooperation to function at 77 GHz with an analog bandwidth of 200 MHz. Even so, no real-time measurements were reported since the OFDM signal was staggered over a few bands with time division multiplexing (TDM) to achieve a 4 GHz total bandwidth, although with this custom SDRS, a similar setup as used in this thesis will certainly be possible.

Most researchers instead perform time-multiplexed measurements only to superimpose them using digital signal processing techniques. This method of measurement do not and cannot take into account the coupling between the antennas and most likely must be done entirely in the anechoic chamber since the free room measurement or environment must remain unchanged for the duration of the measurement. Due to this extremely rigid and controlled measurement environment needed, an outdoor measurement especially with moving targets is not possible and have never been done.

The set up of the USRPs for MIMO Radar use is a challenging task due to many hardware limitations. The analog signal bandwidth is dictated by the Ethernet connection and the host computer's speed (as well as the daughterboard) hence necessitating the use of a super computer (with 64 GB RAM and server-type processor i.e. Intel Xeon) as well as two 10 Gb Ethernet cards, each connected to two 10 Gb cables to the four USRPs. Each USRP has two daughterboards (RF frontend boards), each with one transmit and one receive chain. The Tx port is made so that it can also function as an Rx, hence there is a chip on the daughterboard, which switches the transmit port's signal to the receive chain of the same daughterboard. Although the coupling is about −40 dB, due to the low transmit power, this coupling factor is enough to push the signal to the DAC's clipping/saturation region, thus distorting the received radar signal. For this reason, one USRP is either used as two Tx or Rx sources. As a result, a 4×4 radar configuration will require four USRPs.

A lot of work on the synchronization and optimization has been done in [Bas16] and [Orz16] to enable the massive data throughput[19] from the host computer to the four USRPs simultaneously to achieve the 100 MHz analog bandwidth. Some of these optimizations were made on the signal processing blocks of the open-source software development kit called 'GNU Radio'. These optimizations enabled the real time operation of the MIMO Radar at 100 MHz hence outdoor measurements with a moving target can be performed with the effect of antenna coupling taken into consideration. In this final chapter, the calibration of the USRPs for radar estimation accuracy are discussed and then several indoor and outdoor measurements with verifications are presented. Here only the 2D+velocity concept will be proven since a 4×4 setup for both azimuth and elevation do not yield enough angular resolution for a reasonable estimate. Nevertheless this proof of concept can be extended to the 3D+velocity with more USRPs.

7.1 Calibration of setup

A calibration of the USRPs is used to control the signal start times of all the Tx and Rx as well as to take any initial phase rotations, amplitude imbalance, and frequency offset over the subcarriers into account. To synchronize and stabilize the timing of all Tx and Rx of all USRPs, a common clock and frequency

[19] with 16-bit I and Q symbols, a streaming rate of 3.2 Gbps is needed to generate a 100 MHz analog instantaneous bandwidth. For four transmit streams (host to USRP) and four receive stream (USRP to host), the connection must be able to stream at 12.8 Gbps with full duplex.

reference source is fed to the USRPs directly using the ETTUS OctoClock-G along with same-length but not phase-matched SMA cables. All four cables from the Tx ports are connected to all Rx port cables via two 1:4 power divider/combiner as shown in Fig. 7.1. In this way, the group delay and phase offsets can be accounted for up to the cables, which are also used to connect to the antennas. A Python script for GNU Radio is written to track the clock signals and then lock the phase locked loop (PLL) in phase. This is the initial phase and is different at every power up[20]. If left uncorrected, this initial phase causes a rotation to the constellation diagram as though there is a time delay and distorts the radar range estimation accuracy.

Next the ADCs and DACs are started at the same time as the start of the time stamping. Using empirical methods the group propagation delay for a particular sampling rate used is determined. A higher sampling rate (as well as longer cables) incur a larger propagation delay in terms of 'samples'. These samples are then dropped from the receive signal as a start time calibration. The group propagation delay is constant over USRP initialization cycles and power up.

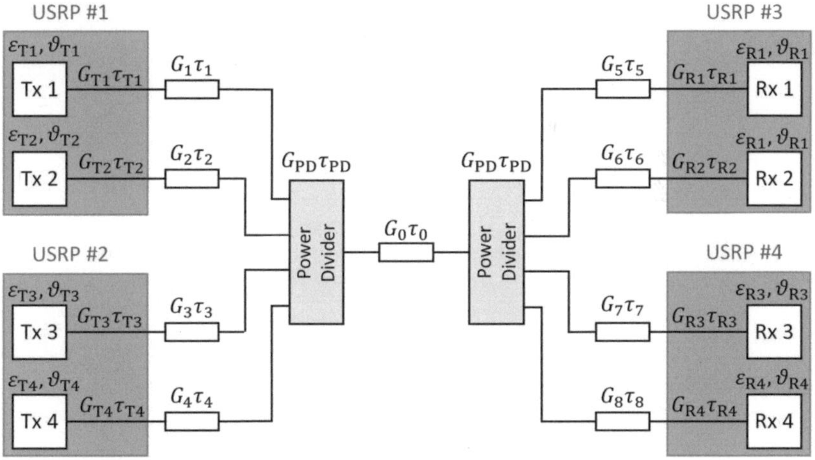

Fig. 7.1: Setup of the USRPs for calibration of internal and external delays (τ), initial phase offset (ϑ), frequency offset (ε) and gain or loss (G).

[20] due to the separate LO used by the Tx and Rx of the same daughterboard, the initial phase is different for both of them.

A calibration matrix $\mathbf{H}_{\text{cal},qp}(n_u,m)$ is made for every Tx-Rx pair with the use of a test OFDM frame for every power up. Fig. 7.1 shows the internal frequency offsets (ε), initial phases (ϑ), group losses ($G_{\text{T/R}}$) and group propagation delays ($\tau_{\text{T/R}}$) as well as the external cable losses (G) and propagation delays (τ) of the entire system setup. Assigning the desired user index to each Tx, transmitting a test OFDM frame (complete with SCA preambles) over the wired connection would yield an expression analogous to Eq. (2.38). Grouping the terms to reflect their effect, the signal at the q-th Rx is expressed as

$$\begin{aligned}\mathbf{Y}_q(n,m) &= \sum_{p=0}^{P-1} \mathbf{X}_p(n_u,m)\, G_{\text{g},qp}\, \mathrm{e}^{\mathrm{j}(\vartheta_{\text{T}p}(n_u)+\vartheta_{\text{R}q}(n_u))} \left(\mathrm{e}^{\mathrm{j}\varphi_{\text{R}_h}(u)}\, \mathrm{e}^{-\mathrm{j}2\pi(iN_{\text{ch}})\Delta f T_{\text{g},qp}}\right)\cdot \\ &\quad \left(\mathrm{e}^{\mathrm{j}\varphi_{\text{train}}}\, \mathrm{e}^{\mathrm{j}2\pi\,\varepsilon_{\text{g},qp}(mT+n_u T_{\text{S}})}\right) + \mathbf{Z}_{\text{dev}}(n,m) \\ &= \sum_{p=0}^{P-1} \mathbf{X}_p(n_u,m)\, G_{\text{g},qp}\, \mathrm{e}^{\mathrm{j}\varphi_{\text{train}}} \cdot \underbrace{\mathrm{e}^{\mathrm{j}\varphi_{\text{R}_h}(u)}}_{\text{DOA}} \cdot \underbrace{\mathrm{e}^{\mathrm{j}(\vartheta_{\text{T}p}(n_u)+\vartheta_{\text{R}q}(n_u))}}_{\text{DOA \& time delay}} \cdot \\ &\quad \underbrace{\mathrm{e}^{-\mathrm{j}2\pi(iN_{\text{ch}})\Delta f T_{\text{g},qp}} \cdot \mathrm{e}^{\mathrm{j}2\pi\,\varepsilon_{\text{g},qp}\, n_u T_{\text{S}}}}_{\text{time delay}} \cdot \underbrace{\mathrm{e}^{\mathrm{j}2\pi\,\varepsilon_{\text{g},qp}\, mT}}_{\text{frequency offset}} + \mathbf{Z}_{\text{dev}}(n,m).\end{aligned} \tag{7.1}$$

where,

- $G_{\text{g},qp}$ is the group net gain of the signal path qp. Each USRP has two adjustable gain blocks - 1) digital amplitude multiplication factor, and 2) analog amplifier (power amplifier at Tx and low noise amplifier at Rx). Both of these are combined into the term $G_{\text{T}p}$ and $G_{\text{R}q}$. The loss of the 1:4 power divider is G_{PD} while cable losses are $G_0,...,G_8$ respectively. For example, for the signal path from Tx 1 to Rx 1, the group net gain (in scalar) is given by

$$G_{\text{g},11} = G_{\text{T}1}\cdot G_1 \cdot G_{\text{PD}} \cdot G_0 \cdot G_{\text{PD}} \cdot G_5 \cdot G_{\text{R}1}. \tag{7.2}$$

- $\mathrm{e}^{\mathrm{j}(\vartheta_{\text{T}p}(n_u)+\vartheta_{\text{R}q}(n_u))}$ is the initial phase offset as locked by the PLL for both the transmit and receive sides. This term is dependent on frequency hence ϑ is written in matrix form with dependency on the subcarriers allocated through the user index n_u.

- $T_{\mathrm{g},qp}$ is the group delay of the internal and external paths traveled by the signal. Every USRP has an internal group delay as denoted by $\tau_{\mathrm{T}p}$ and $\tau_{\mathrm{R}q}$, while the electrical lengths of the cables used also contribute to the signal delay as denoted by $\tau_0,...,\tau_8$ respectively. τ_{PD} is the delay caused by the 1:4 power divider and is minuscule compared to the other delays.
- $\varepsilon_{\mathrm{g},qp}$ is the frequency offset difference between the Tx and Rx due to the different internal LO. The effect is like a Doppler term.
- $\mathbf{Z}_{\mathrm{dev}}(n,m)$ is the noise of the device.

The frequency error $\varepsilon_{\mathrm{g},qp}$ of the synchronization at every power up over numerous test cycles conducted were found to be contained within one frequency offset bin. At the Rx, a fixed amount of samples N_{drop} is cut away from the receive signal based on the relation $T_{\mathrm{g},qp} = N_{\mathrm{drop}} \cdot T_{\mathrm{S}} \cdot T_{\mathrm{g,res},qp}$ to leave only a delay term that is within one range bin $T_{\mathrm{g,res},qp}$. N_{drop} is dependent on the sampling rate (T_{S}) of the USRP and is constant for all paths due to the same-length cables used. The calibration matrix for the qp-th path is

$$\begin{aligned}\mathbf{H}_{\mathrm{cal},qp}(n_u,m) =& \frac{\mathbf{Y}_{qp}(n_u,m)}{\mathbf{X}_p(n_u,m)}\\ =& G_{\mathrm{g},qp}\,\mathrm{e}^{\mathrm{j}\varphi_{\mathrm{train}}} \cdot \mathrm{e}^{\mathrm{j}\varphi_{\mathrm{R}_h}(u)} \cdot \mathrm{e}^{\mathrm{j}(\vartheta_{\mathrm{T}p}(n_u)+\vartheta_{\mathrm{R}q}(n_u))} \cdot\\ & \mathrm{e}^{-\mathrm{j}2\pi(iN_{\mathrm{ch}})\Delta f T_{\mathrm{g,res},qp}} \cdot \mathrm{e}^{\mathrm{j}2\pi\varepsilon_{\mathrm{g},qp}\, n_u T_{\mathrm{S}}} \cdot \mathrm{e}^{\mathrm{j}2\pi\,\varepsilon_{\mathrm{g},qp} mT} + \mathbf{Z}_{\mathrm{dev}}(n,m).\end{aligned} \tag{7.3}$$

The rest of the USRP hardware error is then corrected simply by an element-wise division with $(\widetilde{\mathrm{Y}}_{qp})_{n_u,m} = (\mathrm{Y}_{qp})_{n_u,m}/(\mathrm{H}_{\mathrm{cal}_{qp}})_{n_u,m}$ right before the radar processing. With this calibration, even the cables are taken into account and all relative phase and amplitude over all Tx and Rx ports become zero.

7.2 Mutual interference tolerance measurement

The first measurement done is a wired measurement with two major goals in mind. The first is to determine the available linear range of the USRP for use without running into intermodulation products and ADC/DAC saturation. The second is to verify the interference tolerance chart that was shown in Fig. 4.7. In order to observe this, all propagation and antenna influences must be eliminated leaving only the hardware imperfections, hence the wired instead of over antennas measurement.

The USRPs used in this work is the ETTUS X310 with SBX daughterboards, capable of operating from 0.4 to 4.4 GHz with a maximum instantaneous analog bandwidth of 120 MHz. Due to this very large operating frequency range, the performance of the amplifier blocks is not 'flat' over frequency and the higher their gain values the lower the third-order input intercept point (IIP3) as can be seen from the performance data sheet in [ett20a]. This can be loosely translated to a higher third-order intermodulation product (hence more noise) when a higher gain value is used. The amplitude of this intermodulation product also depends on the amplitude of the signal amplified at that particular gain. Given that the OFDM signal bandwidth used is 100 MHz, this intermodulation product is often located within the signal bandwidth and cannot be detected until it is large enough.

The USRP amplifier gains can be adjusted till a maximum of about 31 dB. A digital amplitude scaling can also be utilized prior to the amplifier but is limited to a factor of 2 (equivalent to 6 dB for the I and Q modulation signals) in this measurement for both the Radar and the Comm. This is to avoid saturating the ADCs and DACs. Taking this into account, the Radar is set to the lowest gain level and amplitude scaling factor possible at the Tx ports while the Comm's signal gain is increased proportionally to induce the desired $\mathrm{SIR_{in}}$. The gain at the Rx ports is the same for all ports and is maintained throughout the measurements. To make the simulation and measurement results approximate each other, certain steps must be taken as follows:

- Using the setup as shown in Fig. 7.1, the OFDM signals with the user index $u \in \{0,1,2,3\}$ are generated at $\mathrm{Tx} \in \{1,2,3,4\}$ respectively and at all Rx, the radar matrix is processed to obtain the $\mathrm{SNR_{out}}$. Based on the observation of all the $\mathrm{SNR_{out}}$, the amplifier gain of the respective USRP ports are adjusted[21] so that the maximum difference of the $\mathrm{SNR_{out}}$ values between all Tx-Rx port combinations is around 3 dB. This step is done to obtain a relative amplitude calibration, whereby all USRP ports transmit and receive almost the same signal amplitude. At the same time, this gain also provides a reasonable $\mathrm{SNR_{out}}$ value. This transmit gain used is thus called the 'initial gain'.

[21] usually only for the Tx ports but if necessary, also the Rx ports, depending on observation

- With the relative amplitude calibration done, the calibration matrix $\mathbf{H}_{\mathrm{cal},qp}(n_u, m)$ is then obtained from each qp-th path.
- With the initial gain values, the Radar and Comm signals are generated at $u = 0$ (at Tx 1) and $u = 1$ (at Tx 2) simultaneously. This signal is then obtained at one of the Rx. The radar processing is then performed and the $\mathrm{SNR}_{\mathrm{out}}$ value is recorded.
- Varying the transmit gain of Tx 2 (Comm) to produce a different $\mathrm{SIR}_{\mathrm{in}}$ and the frequency offset of the Comm signal, the previous step is repeated with all the parameters at Tx 1 and at all receive ports of the USRP maintained. In this way the $\mathrm{SNR}_{\mathrm{out}}$ over various $\mathrm{SIR}_{\mathrm{in}}$ values and frequency offsets can be charted and plotted in a figure.

Plotting the $\mathrm{SNR}_{\mathrm{out}}$ with the range of $\mathrm{SIR}_{\mathrm{in}}$ from 0 to −30 dB over $-1 \leq \delta_w T_0 \leq +1$, the measured as well as the simulated tolerance charts are as shown in Fig. 7.2. The maximum achievable $\mathrm{SNR}_{\mathrm{out}}$ is approximately 65 dB (when $\delta_w T_0 = 0$). To match the measurement conditions the simulation condition was merely scaled with an SNR = 14 dB (the logic behind the value is explained in Section 7.3.1). It is obvious that the measured and simulated tolerance charts correspond well with each other thus verifying the tolerance chart.

The chart is only scaled from a $\mathrm{SNR}_{\mathrm{out}}$ of 10 dB onward since the $\mathrm{SNR}_{\mathrm{out}}$ plateaus[22] at 10 dB. The USRP sampling frequency, signal bandwidth and operating frequency have no effect on the outcome. This has been corroborated with measurements taken with the USRPs at various aforementioned parameters.

Determination of USRP linear operating region

To determine the linear operation region, it is sufficient to compare the simulated results in Fig. 7.2(a) with the measured results in Fig. 7.2(b). Looking at the $\delta_w T_0 = 0$ line of Fig. 7.2(a), it can be seen that with a perfect hardware, the maximum $\mathrm{SNR}_{\mathrm{out}}$ is retained no matter the $\mathrm{SIR}_{\mathrm{in}}$ (i.e. the gain) of the Comm. This value however is not preserved in Fig. 7.2(b) due to the amplifier's non-linearity. At $\mathrm{SIR}_{\mathrm{in}} = -16.7$ dB the $\mathrm{SNR}_{\mathrm{out}}$ decreases to below 60 dB. Hence the gain values used up to this point are considered to be in the 'linear region'.

This gain used for the Comm signal at this $\mathrm{SIR}_{\mathrm{in}}$ value encompass the gain difference between the Radar and the Comm as well as the initial gain value

22 refer to Section 4.3.1

(at Tx, 1 and Tx 2), which is about 3 dB. Hence the maximum gain of the amplifier that can be used in this case is 16.7 + 3 = 19.7 dB. This value was also verified with a spectrum analyzer, whereby a higher gain value used lead to obvious intermodulation products, which can be seen at the edges of the signal bandwidth. In the subsequent sections featuring the interference cancellation measurements, the Comm (interferer) signal will always be generated within this so called linear region. If this principle is not adhered to, the interference

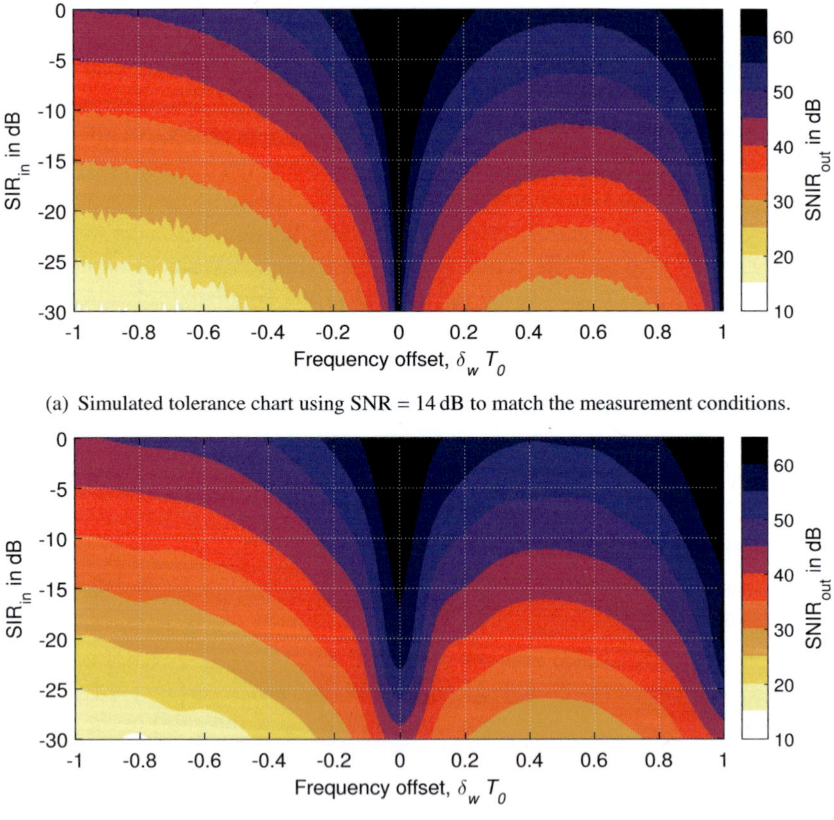

(a) Simulated tolerance chart using SNR = 14 dB to match the measurement conditions.

(b) Measured tolerance chart.

Fig. 7.2: Tolerance chart: Simulated vs. measured charts of the $\mathrm{SNIR}_{\mathrm{out}}$ vs. $\mathrm{SIR}_{\mathrm{in}}$ for one Comm at $-1 \leq \delta_w T_0 \leq +1$. Each level of color represents 5 dB.

cancellation will leave a residue visible in the interference canceled radar and DOA images since the noise caused by the intermodulation products cannot be effectively canceled.

7.3 Case studies measurement verification

The measurement verification performed here is meant to gauge how well the interference cancellation algorithm performs in a real case with an imperfect hardware. In the same manner as was done for the mutual interference tolerance chart in the previous section, the relative amplitude calibration is performed and wired measurements are performed according to the case studies presented in Section 5.6.

To take into account the signal power as transmitted and received at the USRP, the simulation must be set up to emulate the hardware conditions. One of the simplest maneuver is to scale the SNR of the received signal at the receiver of the simulation. The $\mathrm{SNR}_{\mathrm{out}}$ refers to the SNR of the received signal matrix along with the processing gain G_{per} and the loss from the Hamming window[23]. Hence by knowing the maximum $\mathrm{SNR}_{\mathrm{out}}$ at the port used, the SNR can be found. This SNR is then put into the simulations with the single radar target at $r = 0.5\,\mathrm{m}$ with no velocity to emulate the USRP group delay. For a given SNR, the RCS of the single target does not influence the $\mathrm{SNR}_{\mathrm{out}}$ outcome and is set to $1\,\mathrm{m}^2$. Apart from the mentioned effects, no other effects are taken into account in the simulation.

All OFDM parameters except for the f_{c}, Δf and parameters related to T_0 are as shown in Table 5.2. Although the USRP sampling frequency, signal bandwidth and operating frequency have no effect on the outcome, the sampling frequency was chosen to be 10 Msps (equivalent 10 MHz analog bandwidth) to result in $T_0 = 1.024 \cdot 10^{-4}\,\mathrm{ms}$, and the operating frequency is 2.45 GHz. These settings were chosen so as not to push the host computer into overclocking mode all the time, which in turn reduces the probability of an erroneous transmission and reception[24] of the signals.

23 [Bra14, ch. 3, p. 26] offers more detail into the SNR for the periodogram radar processing method

24 overflow and underflow problems as defined in [ett20b]

Since only four USRPs can be controlled and synchronized at the same time with the same host PC, the Radar and Comm measurements are performed separately then linearly added in Matlab. This method has been tested when performing the mutual interference tolerance measurements in the previous section and in [SNB^{+}16] and was found to yield the same outcome. In the following sections the comparison between the simulated and measured results of the SNIR_{out} versus SIR_{in} and frequency offset of the Comm signal(s) for the three case studies will be presented.

7.3.1 Case Study I verification

For this case study, the subcarrier allocation to the 4-Tx Radar and 1-Tx Comm is as depicted in Fig. 7.3(a). The Radar is assigned $u \in \{0,1,2,3\}$ generated at $\text{Tx} \in \{1,2,3,4\}$ respectively. The Comm is assigned $u \in \{5\}$ and is generated at $\text{Tx} \in \{1\}$. The SNIR_{out} is processed at all Rx for all Tx, resulting in 16 values. It has been seen in Section 5.9.1 that the subchannels belonging to $u = 3$ experience the worst isolation hence the SNIR_{out} versus SIR_{in} figure will be plotted based on the subchannels of $u = 3$ at Rx 1.

The maximum SNR_{out} of the measured radar image at Rx 1 is $\text{SNR}_{\text{out}} \approx 62\,\text{dB}$ without any interference at $\text{SIR}_{\text{in}} = 0\,\text{dB}$ and $\delta_w T_0 = 0$. As have been mentioned, this value encompasses the Radar's received signal along with the cable losses, amplifier gains, system noise figure, SNR, G_{per} and windowing loss. The processing gain is $G_{\text{per}} = 45.15\,\text{dB}$ according to Eq. (2.48) for $N_{\text{ch}} = 8$ configuration. The Hamming windowing loss[25] is taken to be 2.7 dB. Hence the actual power level of the signal is $62 - 45.15 - 2.7 \approx 14\,\text{dB}$, which is then set as the SNR at the Rx in the simulation.

Shown in Fig. 7.3(b) and Fig. 7.3(c) are the simulated and measured SNIR_{out} versus SIR_{in} plots respectively. The dashed lines represent the SNIR_{out} of the Radar signal with interference over the frequency offsets of 0% to 30% (denoted as f0, f1, f3, f7, f15, f30 respectively) of the subcarrier spacing and the solid lines are the interference canceled results with legends marked with a '#'. Both of the figures correspond well with each other with minor differences.

It can be seen that the maximum SNIR_{out} of the simulated results are slightly higher by around 3 dB and that the simulated Radar signal with interference saturates at the SNIR_{out} of 38 dB at f30 compared to the measured equivalent,

[25] refer to Chapter 2.5.5 for windowing losses.

which goes down linearly to 30 dB without signs of plateauing. This plateauing phenomenon can be attributed to the very low SNR used in the simulations, which causes a faux elevation of the noise floor level. However, with this simple 1-Tx Comm (interferer) configuration, the point of the jump that leads to the steady improvement after interference cancellation is at the same SIR_{in} values for both the f15 and f30 curves at $SIR_{in} = -6$ dB and -11 dB respectively .

From $SIR_{in} = -17$ dB onward to the more negative values, the measurement results droop into a slight curve, which is expected since the linear operating region has been exceeded. With this, all interference canceled signals will also obtain a maximum value that is dictated by this downward curve.

7.3.2 Case Study II verification

The measurement for Case Study II is done in the same way as for Case Study I with the exception that four transmitters are used at the Comm node with the subcarrier allocation as shown in Fig. 7.4(a). Here the initial gain values over all four transmitters range from 3 to 6 dB to achieve the approximately the same SNR_{out} at all transmit-receive ports. With the addition of three Comm transmit signals generated, the level of noise can be expected to rise by at least four times or 6 dB. Due to the imperfect hardware, a different gain setting incurs a different level of noise floor and this overall rise in the noise floor is foreseen to also affect the range of the linear operating region.

Fig. 7.4(b) shows the simulated results of Case Study II over various SIR_{in} and frequency offset of Comm's signals. The results were taken at Rx 1 at the subchannels of $u = 3$. In the same way as Case Study I, the SNR has been set to 14 dB in the simulation. All the curves then exhibit a plateauing effect at around $SNIR_{out} = 32$ dB (6 dB higher than in Case Study I) due this this faux noise floor added at the Rx.

The measured equivalent is as shown in Fig. 7.4(c). Here, the interaction between the Comm and Radar signals are more complicated due to the 3 dB imbalance (of the initial gains) of the signal amplitude and noise floor at every USRP transmit and receive ports. This causes the four Comm signals to have a slightly different power and noise floor level from one another. The 6 dB rise in the noise floor due to the four different margins of the intermodulation product at different gain values will definitely cause the $SNIR_{out}$ of the f0 curve to start drooping at about 6 dB before $SIR_{in} = -17$ dB as with the Case Study I.

Furthermore, with the 3 dB difference in the initial gains, this further adds a proportional amount of intermodulation product and decreases the linearity of the f0 curve over SIR_{in} by a further 3 dB. Hence the f0 curve is expected to start declining at $SIR_{in} \approx -17+6+3 = -8$ dB, which is the case in Fig. 7.4(c). Due to the complications detailed above as well, the f15 curve did not improve by much after interference cancellation until $SIR_{in} = -5$ dB as compared to the simulated equivalent at $SIR_{in} = -2$ dB. The f30 however managed to match the simulated curve.

Taking the above discussed factors into considerations, it can be seen that the trend of the curves with interference as well as the ones after interference cancellation match well with the simulated results.

7.3.3 Case Study III verification

In Case study III, all subchannels are used by both the Radar and the Comm as shown in Fig. 7.5(a). Compared to Case Study I and II, the Comm signal now occupies all channels that are also occupied by the Radar hence the increase of the noise level as seen by the Radar's channel is also eight times (9 dB) more than in the previous case studies. As such the SNR at the Rx is set to $14-9 = 5$ dB in the simulations.

The simulated and measured results are as shown in Fig. 7.5(b) and Fig. 7.5(c). While all the curves from f0 to f30 approximate each other in the simulations, the curves in the measurement results differ slightly with each other. The difference between the simulation and measurement results is especially obvious when comparing the f0 curve as well as the plateauing point at $SIR_{in} = -21$ dB. This can be attributed to the imbalance in the USRP ports as have been discussed in Case Study II, which might have a slightly better SNR than what have been assumed for the simulations.

Case Study measurement conclusion

All things considered, the measurement verification of all the case studies presented approximate the simulated values very well and hence can be concluded as reliable results. This proves that the interference cancellation algorithm also works with imperfect real signals, which suffer from typical non-linear distortions, albeit with a slightly reduced performance.

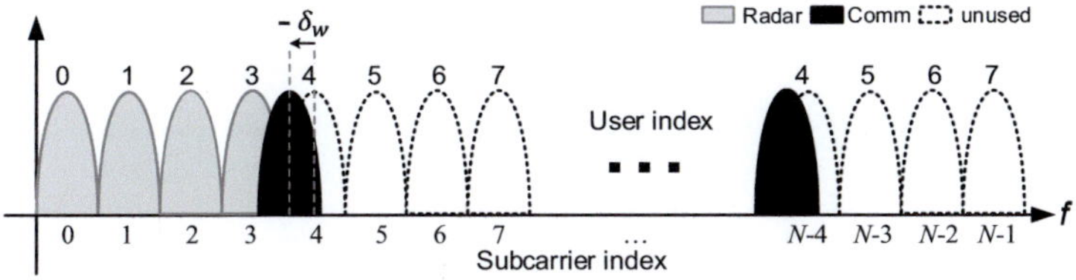

(a) Subcarrier assignment for the Radar $u = \{0, 1, 2, 3\}$ and Comm $u = 4$.

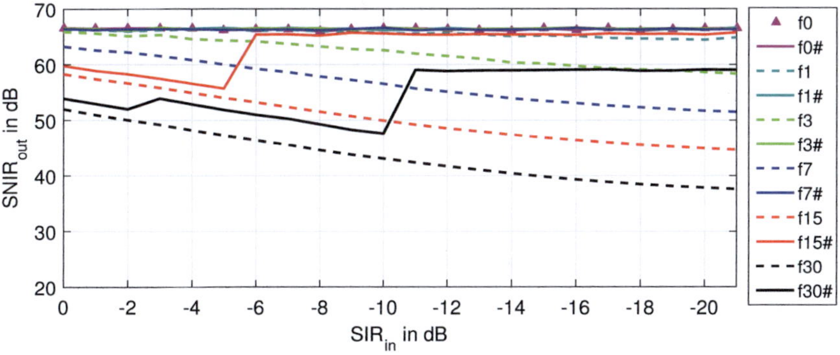

(b) Simulated with SNR = 14 dB.

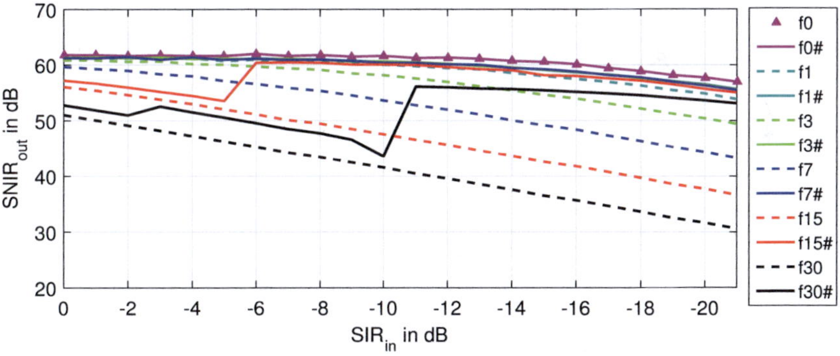

(c) Measured.

Fig. 7.3: Case Study I: Simulated vs. measured curves of SNIR_{out} vs. SIR_{in} for one interfering communication partner (Comm) node with one transmitter at various frequency offsets. The frequency offsets of $\delta_w T_0 = \{0, -0.03, -0.07, -0.15, -0.3\}$ are denoted by f0, f3, f7, f15, f30. Legends marked with '#' denote the interference canceled equivalent.

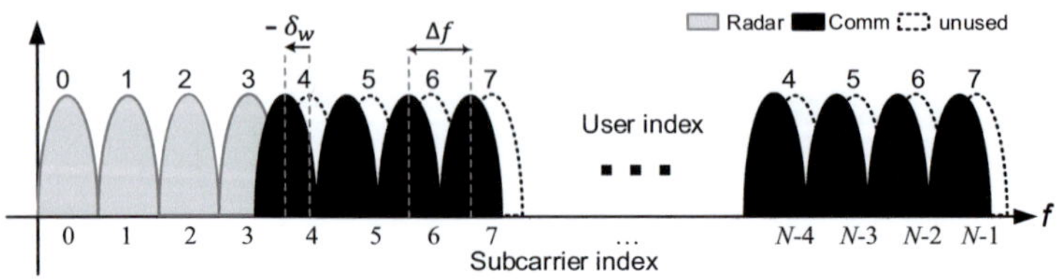

(a) Subcarrier assignment for the Radar $u = \{0,1,2,3\}$ and Comm $u = \{4,5,6,7\}$.

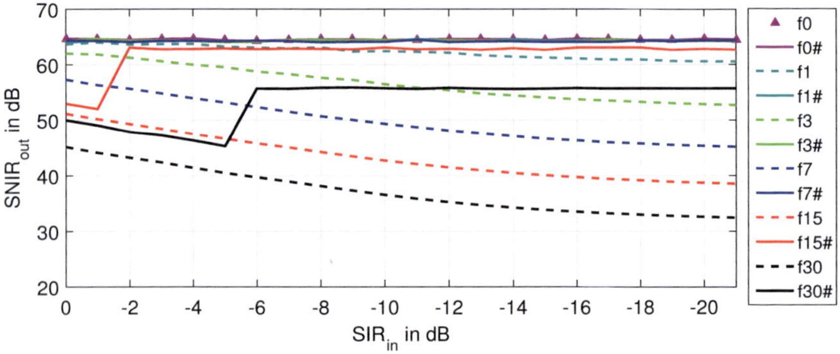

(b) Simulated with SNR = 14 dB.

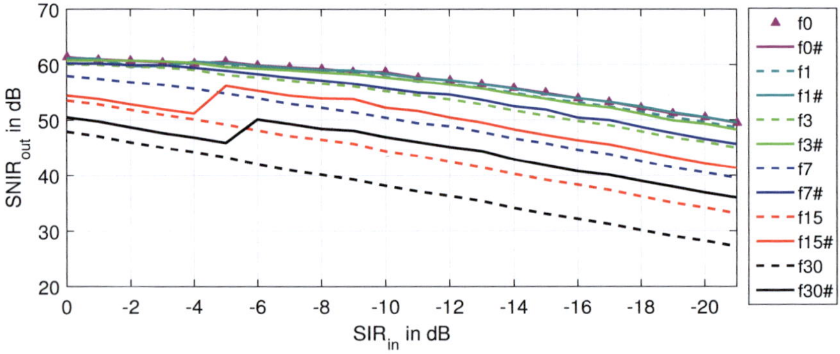

(c) Measured.

Fig. 7.4: Case Study II: Simulated vs. measured curves of SNIR_{out} vs. SIR_{in} for a 4-Tx interfering communication partner (Comm) node at various frequency offsets. The frequency offsets of $\delta_w T_0 = \{0, -0.03, -0.07, -0.15, -0.3\}$ are denoted by f0, f3, f7, f15, f30. Legends marked with '#' denote the interference canceled equivalent.

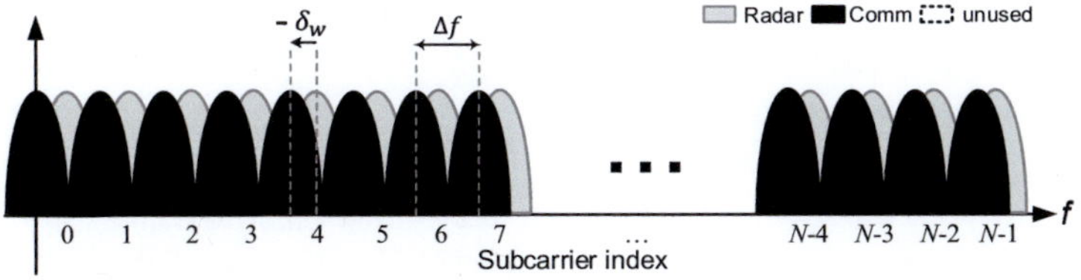

(a) Subcarrier assignment for the Radar $u = \{0\}$ and Comm $u = \{0\}$.

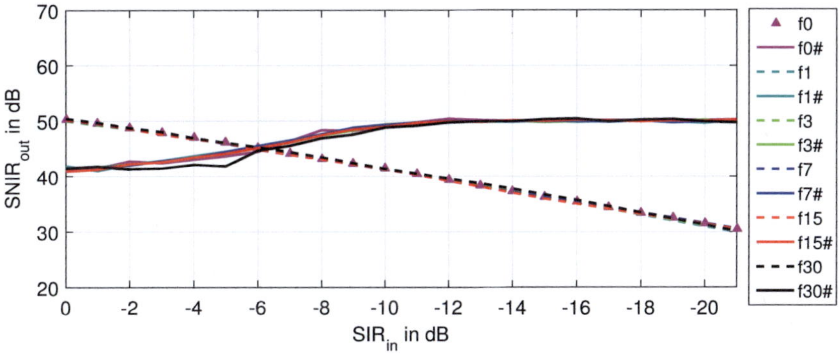

(b) Simulated with SNR = 5 dB.

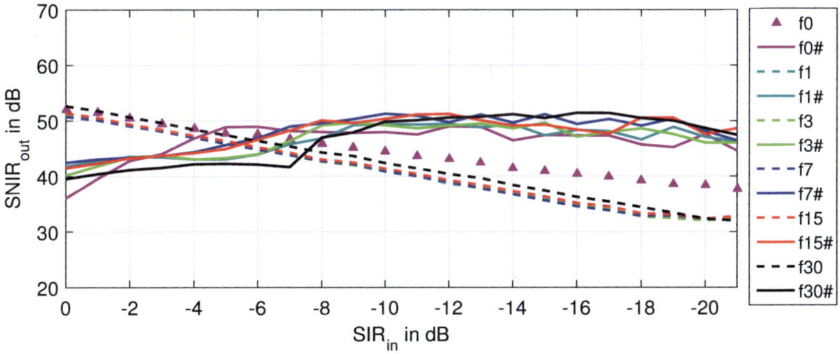

(c) Measured.

Fig. 7.5: Case Study III: Simulated vs. measured curves of $\mathrm{SNIR}_{\mathrm{out}}$ vs. $\mathrm{SIR}_{\mathrm{in}}$ for one interference communication partner (Comm) node at various frequency offsets. The frequency offsets of $\delta_w T_0 = \{0, -0.03, -0.07, -0.15, -0.3\}$ are denoted by f0, f3, f7, f15, f30. Legends marked with '#' signifies the interference canceled equivalent.

7.4 Measurements with simple patch antenna arrays

To start off with the wireless measurements, simple patch antennas are designed at the operating frequency of $f_c = 4.05\,\text{GHz}$. The size of the individual antenna element is 4 cm at the E-plane and 4.4 cm at the H-plane dimensions as shown in Fig. 7.6(a). The receive antenna array is fabricated with four patch elements on the same substrate spaced 5 cm apart as shown in Fig. 7.6(b). This results in a $d_R = 0.675\,\lambda_c$. The transmit antenna elements are fabricated individually and they are spaced $d_T = 4 \cdot 0.675\,\lambda_c = 20\,\text{cm}$ apart. Both the transmit and receive arrays are mounted on 3D printed plastic frames that do not interferer with their radiation patterns as shown in Fig. 7.6.

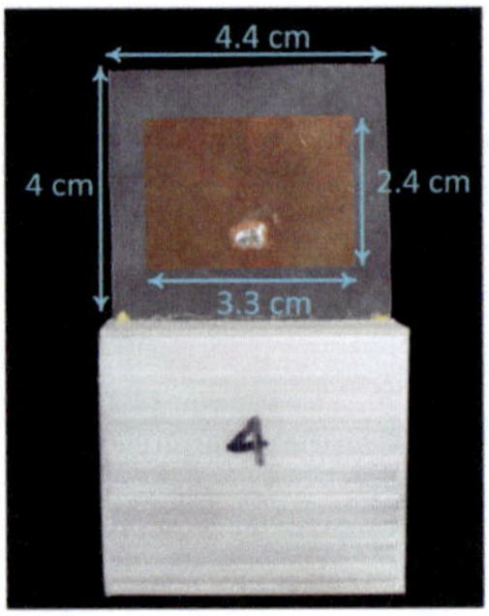

(a) Individual Tx element.

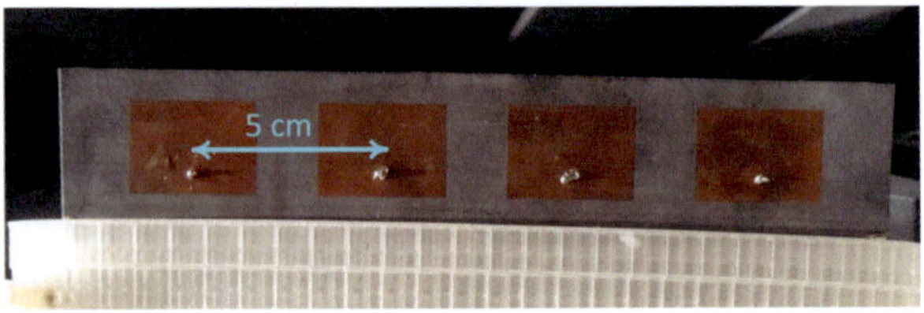

(b) Rx array.

Fig. 7.6: Patch antenna element for the transmitter and array for the receiver.

Fig. 7.7 shows the arrangement of the physical antenna arrays to result in a 16-element virtual array. The receive array is placed above the Tx array and is separated by about $d_{TR} = 5\,\lambda_c$. This array spacing does not affect the azimuth DOA estimation results. The antennas arrays' angular ambiguity plot is shown in Fig. 7.8(a). Due to the larger than optimal element spacing, high sidelobes appear at around $\phi = [-30°, 30°]$, thus shrinking the unambiguous azimuth FOV shrinks to around the aforementioned azimuth angles. Taking into account the element factor in Fig. 7.8(b) , the virtual array radiation pattern is as shown in Fig. 7.8(c). Theoretically, the angular resolution based on the HPBW (Eq. (3.42)) is 5.3°.

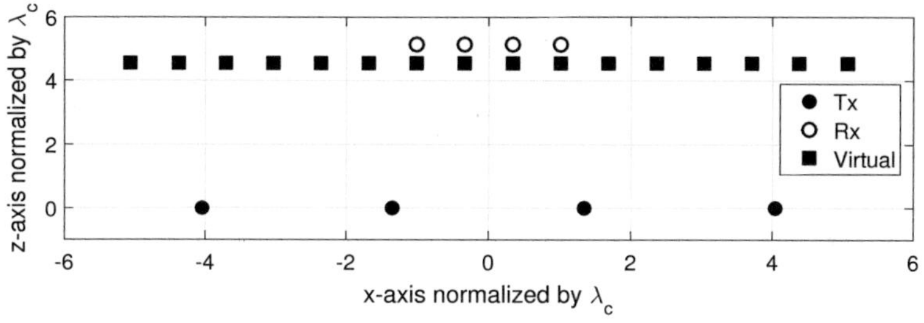

Fig. 7.7: Position of Tx, Rx and virtual array elements of the patch antenna arrays.

Three different measurement scenarios for the verification of the 2D+velocity capabilities of the MIMO RadCom are presented in this thesis as follows:

- Chamber: EMC chamber verification measurements of static targets with Doppler emulator.
  - Azimuth angle resolution
  - Elevation
  - 2D+velocity
  - Interference cancellation from a 4-Tx Comm node (with simulation verification)
- Outdoor 1: Outdoor DOA measurement of static targets.
- Outdoor 2: Outdoor measurement of a moving target.
  - Interference cancellation from a 1-Tx Comm node

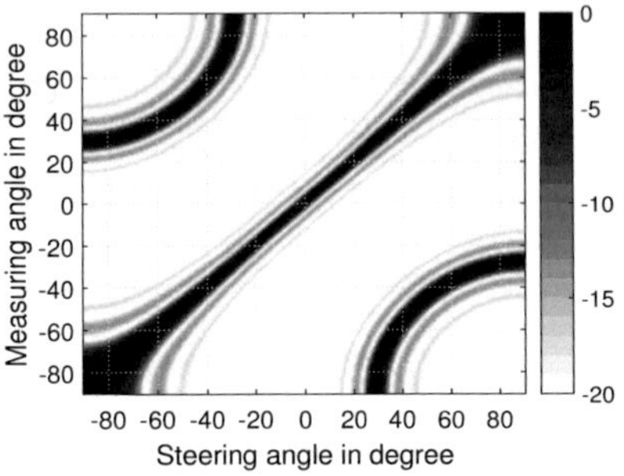

(a) Simulated angular ambiguity plot.

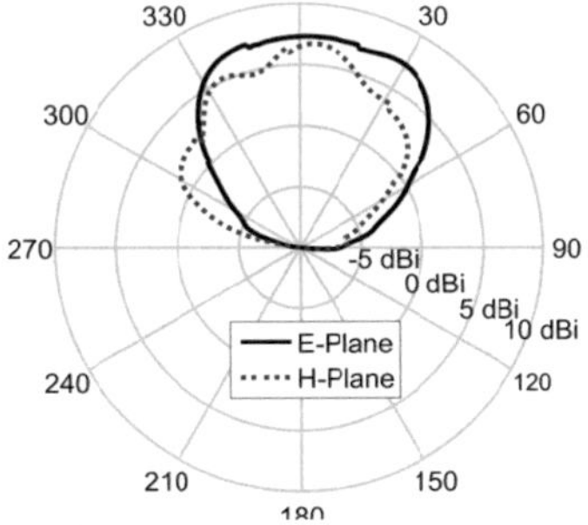

(b) Measured patch antenna element radiation pattern in the main direction.

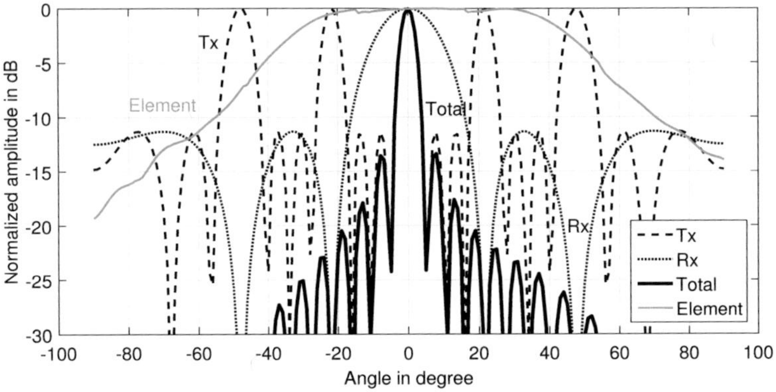

(c) Resulting radiation pattern (Total) from simulated array factor based on the position of the patch antennas (Tx, Rx) with measured E-plane element factor (Element).

Fig. 7.8: Physical 4×4 1D transmit and receive arrays with their element radiation pattern and resulting virtual radiation pattern characteristics. The unambiguous angular FOV is about $\phi = [-30°, 30°]$ from 7.8(a).

7.4.1 Measurement setup and parameters

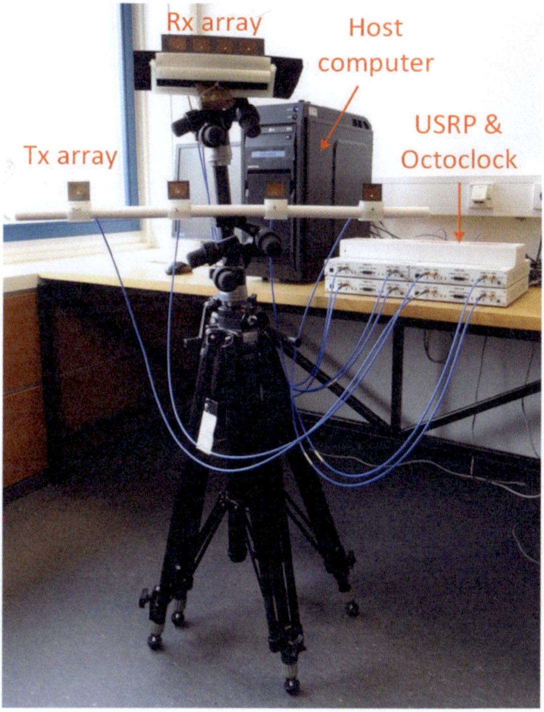

Fig. 7.9: Setup of the four USRPs with Octoclock-G, host computer and antenna arrays for wireless measurements. RF absorbers not shown here.

The measurements are done using the parameters in Table 7.1 unless mentioned otherwise. The transmit and receive array indices are according to the order $p = q = \{0, 1, 2, 3\}$ located on the x-axis. There are $N_{\text{ch}} = 8$ channels and the user index $u = \{0, 1, 2, 3\}$ is assigned to each p-th antenna of the Radar in a regular order. All radar and DOA images are shown with Hamming windowing.

The setup of the USRPs, host computer and antenna arrays are as shown in Fig. 7.9 with the ordering of spacings of the antenna arrays as depicted in Fig. 7.10. Since these antenna arrays are mounted on metal tripods, RF absorbers are placed between these arrays to minimize direct coupling. RF absorbers are also placed at the back of the arrays as well as at the legs of the tripods to shield them and minimize undesired reflections.

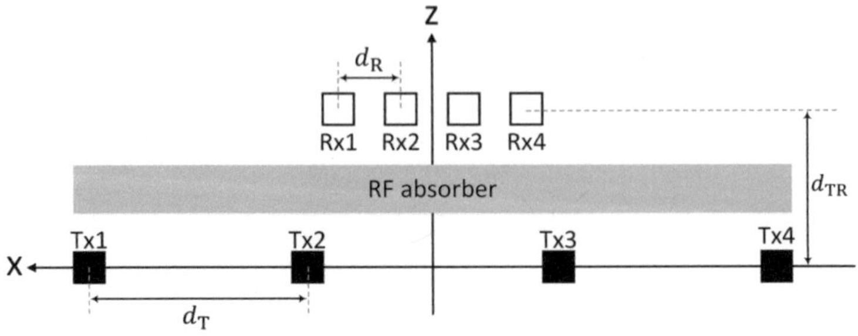

Fig. 7.10: Antenna array setup for measurement. $d_T = 20\,\text{cm}$, $d_R = 5\,\text{cm}$, $d_{TR} = 38\,\text{cm}$.

Table 7.1: Radar parameters on USRPs.

Symbol	**Parameter**	**Value**
f_c	Carrier frequency	4.05 GHz
N	Number of subcarriers	1024
M	Number of subsequent symbols	256
Δf	Subcarrier spacing	97.66 kHz
T_0	Elementary OFDM symbol duration	10.24 µs
T_{CP}	Cyclic prefix duration	1.28 µs
T	Total duration of one transmit OFDM symbol	11.52 µs
BW	Total signal bandwidth	100 MHz
Δr	Range resolution	1.5 m

Three types of trihedral corner reflectors are used as summarized in Table 7.2. These RCS values represent the maximum possible theoretical values but the actual RCS values are dependent on their orientation with respect to the incident rays of the Radar. In order to obtain a backscattering toward the Radar, the incident rays must be directed to the effective areas[26] of the reflectors. According to Robertson in [Rob47], assuming that the rays illuminate the trihedral reflector's effective area, four cases can happen, namely,

1. if the angle of the incident ray is too oblique (neither parallel nor right angle), the ray may not return to the Radar.

[26] [Rob47] and [Kno12] give definitions of RCS and effective areas of various reflectors.

2. if the ray is exactly perpendicular to one of the three trihedral's planes, the ray will be reflected only once and returned to the Radar.
3. if the ray is exactly parallel to one of the three planes, the ray will be reflected twice (like in a dihedral reflector) and returned to the Radar.
4. in any other angles between the aforementioned ones, a triple bounce from all three planes of the reflector will occur before the ray is returned to the Radar.

Table 7.2: Radar target RCS used in the measurements.

Target	Reflector type	Max. RCS in m^2	Max. RCS in dBm^2
A	trihedral	47.7	16.8
B	multifaceted trihedral	2.4	3.9
C	trihedral	1.2	0.9

7.5 Measurement verifications in the EMC chamber

The electromagnetic compatibility (EMC) chamber has the area of around 11 m (length) × 5 m (width) × 6 m (height) and is padded with anechoic foams throughout the walls and ceilings except the floor. The floor is layered with metal sheets. To avoid secondary reflections to the Radar, which arise from the corner reflectors to the metal flooring then back to the Radar, RF absorbers have been strategically placed and numerous radar measurements were done to ensure that the radar and DOA images show only the desired reflections from the targets. These strategies include placing more RF absorbers in the direct reflection path of the reflectors, tilting the reflectors upwards to face away from the metal floor, etc. No external power amplifiers are used since the maximum range that can be estimated is limited by the length of the chamber at 11 m. The internal amplifiers of the USRPs are sufficient to provide a radar signal that can illuminate up to 80 m outdoors.

Since the length of the room is only 11 m, the possible placements of the targets are simulated according to the contiguously distributed user index to gain an idea of what can be expected to emerge from the DOA results. A single object is placed at $r = \{1, 2, 3.5, 6, 9\}$ m away from the Radar and the DOA image's azimuth cut is plotted with the results shown in Fig. 7.11. Any target situated at 1 to 2 m away would emerge with multiple peaks due to the antenna near-field defocusing effect (see Eq. (3.30)). Hence care has been taken to place the nearest reflector at more than 3 m away.

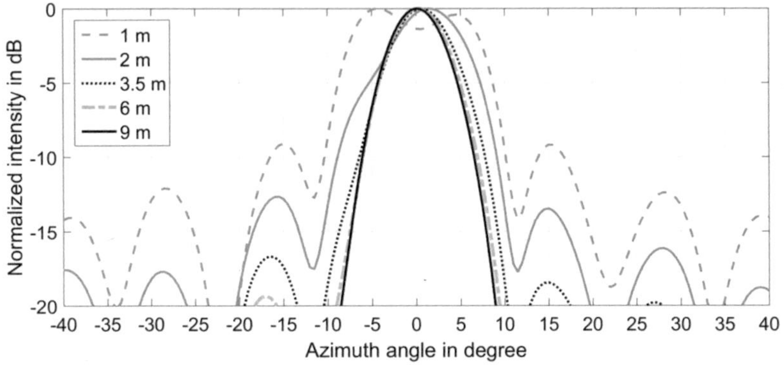

Fig. 7.11: Chamber: Expected DOA results (with Hamming windowing) based on simulations of a target at $r = \{1, 2, 3.5, 6, 9\}$ m in the EMC chamber using the Fourier Beamforming method. Targets that are too near the radar suffer from a defocusing effect (see Eq. (3.30)).

Near-range clutter with specular and diffuse reflections

The effect of near-range clutter due to the metal floor, diffuse reflections from the RF absorbers and metal tripods on the radar images is also investigated. A simple scenario is chosen to model these near-range clutter along with a target. The near-range clutter can be bundled and modeled as reflection points as shown in Table 7.3. The far-field region of the antenna array is estimated as $2 \cdot (d_R Q)^2 / \lambda_c = 1.08$ m hence the distance for the near-range clutter is set as 1 m to approximate the effect.

The expected influence of the near-range clutter on the DOA image is as shown in Fig. 7.12. It can be seen that the multiple peaks of the reflections at such a near distance add up to form a strong image in the DOA image that spreads over the entire azimuth FOV.

These values along with Target C at 6 m are then put into a complete simulation to obtain the radar image DOA image as shown in Fig. 7.13. The effect of

Table 7.3: Chamber: Simulation parameters for specular and diffuse reflections from near-range clutter.

Target type	Distance in m	Azimuth angle in °	RCS in m^2
near-range clutter	1	−15	$0.5 \cdot 10^{-3}$
near-range clutter	1	1	$0.5 \cdot 10^{-3}$
near-range clutter	1	11	$0.5 \cdot 10^{-3}$
Target C	6	12	1.2

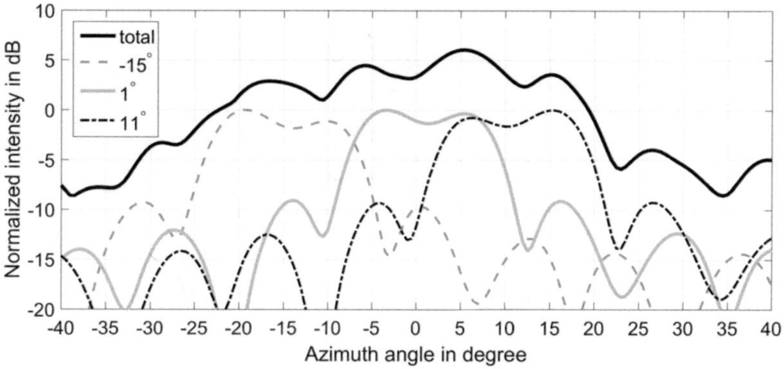

Fig. 7.12: Chamber: Expected influence of near-range clutter e.g. reflections from the metal floor directly in front of the antenna arrays, on the DOA results based on simulations. The values of the reflection points are shown in Table 7.3. The reflections add up to form the 'total' reflection of higher power.

the near-range clutter tend to show up within the 0 to 2 m range and distorts the shape of the target on the DOA images. Numerous test measurements conducted in the same environments confirmed that even with RF absorbers in the vicinity of the antenna array, these near-range clutter still occur. Hence in all the radar estimation results, the range from 0 to 2 m is neglected.

Transmit to receive antenna array coupling

The coupling of the transmit antenna elements with each other were measured with a Vector Network Analyzer (VNA). It has been found that directly neighboring elements (spaced $d_{\mathrm{T}} = 20\,\mathrm{cm}$ apart) have a coupling of −44 dB. Elements that are spaced 40 cm apart have a coupling of −47 dB, while the largest element spacing of 60 cm contribute to a −50 dB coupling effect. These were then modeled in the simulation for the Radar's transmit array. Each antenna element would be coupled with the corresponding values of the other three transmit antenna elements before they are put through the channel simulation.

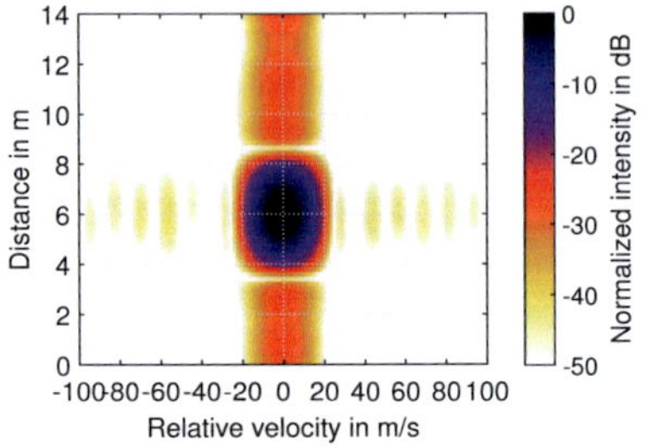

(a) Radar image.

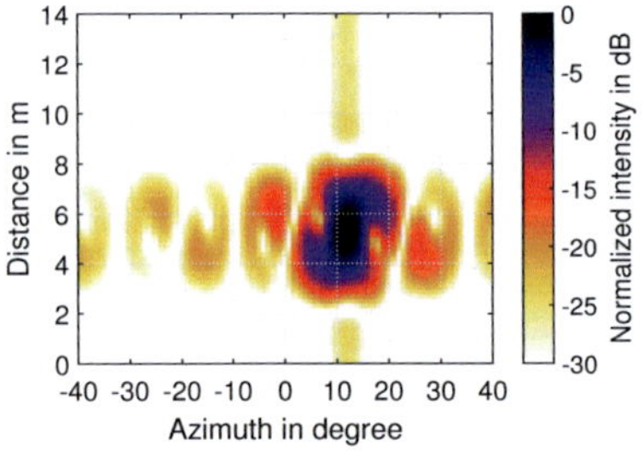

(b) DOA (slice): no windowing.

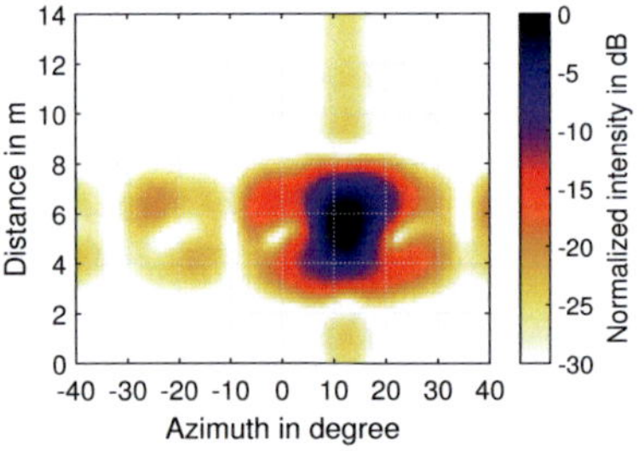

(c) DOA (slice): with Hamming window.

Fig. 7.13: Chamber: Simulation results of the radar image and DOA image slices taken at 0 m/s according to near-range clutter in Fig. 7.12. The specular and diffuse reflections from near-range clutter will appear over the whole angular field-of-view and distort the shape of the target in the DOA images.

The direct coupling of the transmit array with the receive array for the Radar is at the best case, around −83 dB, a value also substantiated with VNA measurement, with several RF absorbers put between the transmit and receive arrays. Referring to Fig. 7.10 there is a coupling of −80 dB for Tx 1 to Rx 1, the best case happens for Tx 1 to Rx 4 with the coupling factor of −83 dB. The coupling between the transmit antennas in the middle i.e. Tx 2 and Tx 3 to all the receive antenna elements is −74 dB. This coupling effect is added directly to the Radar's receive signal according to its coupling factor.

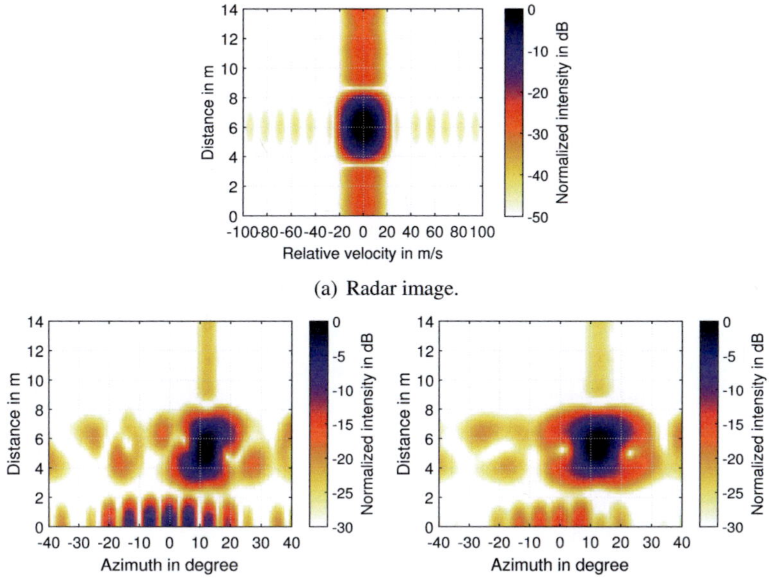

(a) Radar image.

(b) DOA (slice): no windowing.

(c) DOA (slice): with Hamming window.

Fig. 7.14: Chamber: Simulation results of the radar image and DOA image slices taken at 0 m/s showing the effect of transmit to receive antenna array coupling. The effect is mostly to the 0 to 2 m range and distorts the shape of the target in the DOA images.

In this simulation, only Target C in Table 7.3 is used to show the effect of the transmit to receive array coupling. The resulting radar images are as shown in Fig. 7.14. When there is no windowing used, the effects of the coupling are obvious and like the near-range clutter, they are mostly confined to the ranges of 0 to 2 m and distorts the shape of the target in the DOA images. Although this coupling is weak, is still able to add a few decibel of noise into the Radar's receive signal.

The near-range clutter and transmit-to-receive antenna coupling effects will be apparent in all the measured radar and DOA images shown in the following sections. Section 7.5.4 shows the interference cancellation measurement from a 4-Tx Comm (interferer) node with a simulation of the scenario also set up for verification purposes. All the aforementioned effects are taken into account in the simulation, without which, the $SNIR_{out}$ values obtained will differ by more than 6 dB from thr measured equivalent even if the signal power have been scaled to match the measurement conditions.

7.5.1 Chamber: Azimuth angle resolution measurement

In this section, the verification of the azimuth angular resolution using the aforementioned patch antenna array setup will be verified. Fig. 7.15 shows the setup of the two equal-sized small trihedral corner reflectors (Target C). Their Euclidean distance measured from the center of the antenna array with a laser rangefinder is about 5 m. These trihedrals were placed without any tilting to avoid secondary reflections from the metal floor. At the spacing of 75 cm apart, the physical angle between them calculated with simple trigonometry is about 8.5° to the center of the antenna arrays.

Fig. 7.16 shows the radar image of the measurement setup along with the DOA image with and without Hamming windowing. In the figures, due to the coupling of the transmit-receive antennas and some reflection from the metal floor, there exists some clutter at the 0 to 2 m region as expected and is disregarded. To omit showing the effect of other clutters in the vicinity on the DOA image, the color axis is scaled to $[-15, 0]$.

Despite the 24 dB target peak gain (see Section 4.3.1), the noise floor is elevated considerably due to reflections from any reflective surfaces of non-targets (clutter) especially at zero Doppler causing a non-Gaussian behavior. Hence the net gain of the target peak relative to the elevated noise floor becomes approximately ≤ 0 dB, and it is not apparent when comparing to the $\mathrm{SNR}_{\mathrm{out}}$ of the radar image by visual comparison. This phenomena is apparent in all subsequent measurement results.

This setup represents the minimum spacing that allows the two reflectors to be distinguished in the DOA image. Without windowing, the estimated angle between them is 7.5° and with windowing, 9.5°. These outcomes correspond well with the calculated physical angle of approximately 8.5° and hence can be confirmed as a reliable estimate and is taken as the angular resolution. Comparing this value with the theoretical HPBW value of 5.3°, it can be seen that this is also only a small expected difference.

Table 7.4 shows the results of various angular measurements with the reflectors spaced over varying distances apart. ‘Spacing’ refers to the space between the trihedrals, ‘Distance’ is their Euclidean distance from the center of the

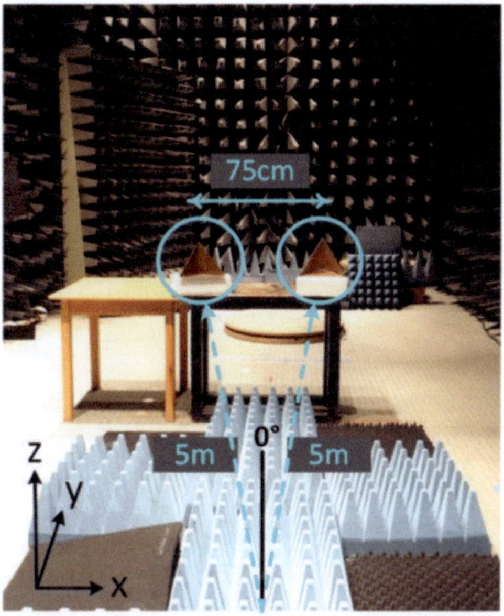

(a) Azimuth resolution measurement scenario.

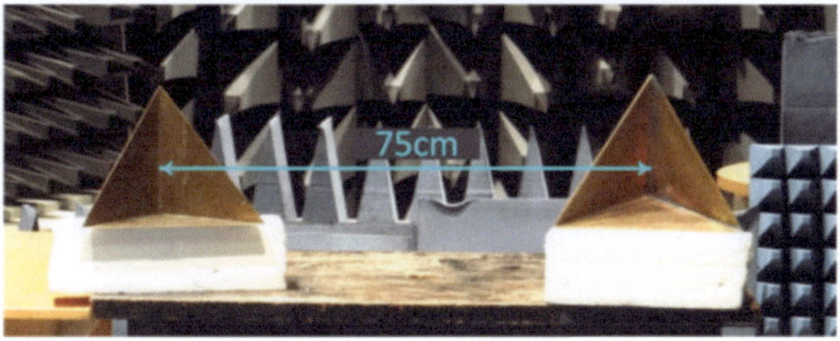

(b) Close up of the trihedral reflectors, Target C.

Fig. 7.15: Chamber: Azimuth resolution measurement setup.

antenna array, ϕ_{phy} is the physical angle calculated with trigonometry, $\hat{\phi}_{\text{DOA}}$ and $\hat{\phi}_{\text{DOA,win}}$ are the estimated angles taken from the DOA image without and with Hamming windowing respectively. It is obvious that the DOA estimated angles correspond well with the physical angles. This is especially true for spacings larger than the angular resolution, since the effect of the lobes of the virtual antenna radiation pattern and windowing no longer impose a large influence. With this, the virtual antenna theory presented in Chapter 3 has been verified to be correct and reliable.

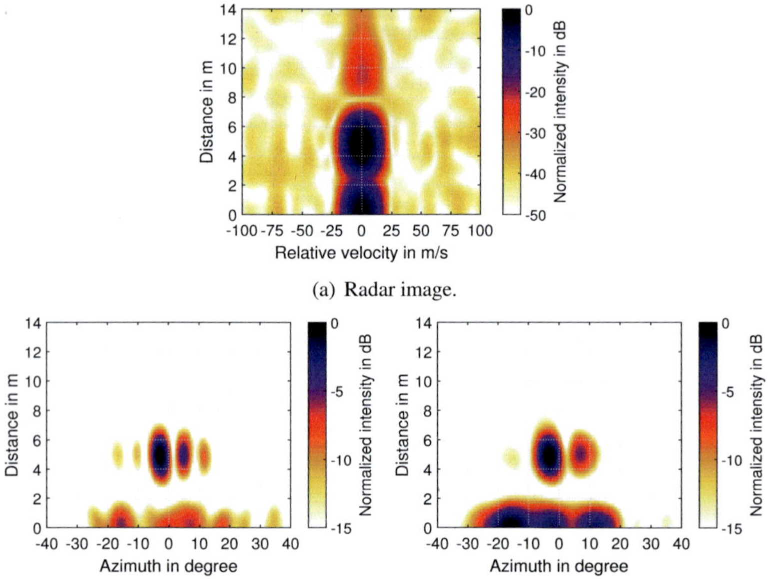

(a) Radar image.

(b) DOA (slice): no windowing.

(c) DOA (slice): with Hamming window.

Fig. 7.16: Chamber: Azimuth resolution radar image and DOA image slices taken at 0 m/s according to setup in Fig. 7.15. The measured angular resolution is approximately 8.5°.

Table 7.4: Chamber: Angle resolution measurements.

Spacing in m	Distance in m	ϕ_{phy} in °	$\hat{\phi}_{\mathrm{DOA}}$ in °	$\hat{\phi}_{\mathrm{DOA,win}}$ in °
0.75	5	8.5	7.5	9.5
1.25	5	14.3	14.0	14.5
1.50	5	17.0	17.5	16.0
0.65	6	6.2	7.0	8.5
0.75	6	7.1	7.0	7.5
1.75	6	16	14.5	15.6
2.35	6	21.4	21.0	21

7.5.2 Chamber: Elevation only DOA measurements

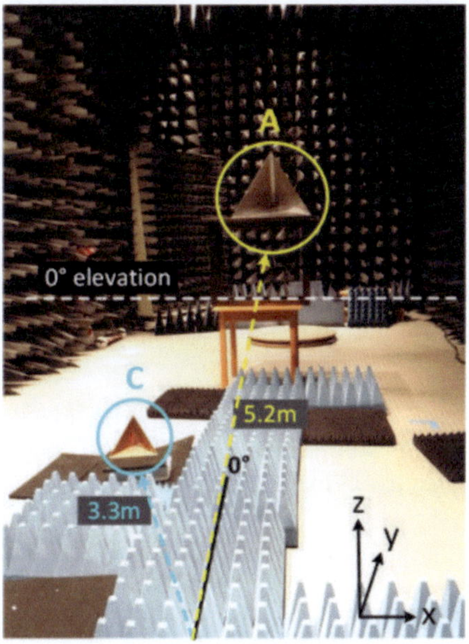

Fig. 7.17: Chamber: Elevation DOA measurement setup.

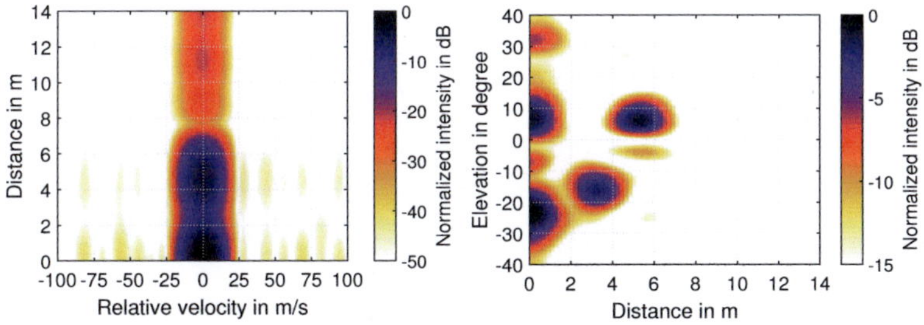

(a) Radar image.

(b) DOA image (slice at 0 m/s).

Fig. 7.18: Chamber: DOA elevation measurement according to setup in Fig. 7.17.

This measurement is conducted only to prove that the elevation plane DOA measurements can be done just like for the azimuth DOA measurements. For this, the transmit and receive antenna arrays are lined up vertically parallel to the z-axis. Again, the d_{TR} will not influence the DOA estimates. The transmit and receive arrays are however spaced adequately apart for the RF absorbers to go in between them to minimize the transmit-to-receive coupling.

Two targets are set up as shown in Fig. 7.17. The height at center of the antenna array to the floor is about 1.3 m and this height is the $\theta = 0°$ elevation point. Target A is elevated on wooden tables to the height of about 2 m from the floor or 0.7 m from the 0° elevation point, making its elevation angle around 7.6°. Target C on the other hand is put on the floor, above some layers of RF absorbers at the negative elevation level of around −20°. Using the laser rangefinder, the Euclidean distances of Target A and C are around 5.2 m and 3.3 m respectively from the center of the antenna arrays.

The measured results are shown in Fig. 7.18, where Target A is estimated to be at 5.25 m, 6°, while Target C is at 3.38 m, −16.5°. The small discrepancy between the physical and estimated angles of Target C is still within the angular resolution and can arise from imperfect physical measurements.

7.5.3 Chamber: 2D+velocity measurements

In this measurement, the capability of the radar to detect velocity (in the 2D+ velocity concept) is to be verified. An active target featuring a Doppler emulator consisting of two signal generators and antennas along with a mixer circuitry as shown in Fig. 7.19 is set up in a scenario shown in Fig. 7.20. The antennas used are the ETS-Lindgren double-ridged horn antennas [dou30].

In order to generate a Doppler frequency f_D, the incident RF_{in} radar signal is down-converted with an LO_1 signal to produce $IF_{in} = LO_1 - RF_{in}$. This signal is then passed through two lowpass filters to remove unwanted sidebands and then amplified with an intermediate frequency (IF) amplifier. The output signal with added Doppler is then $RF_{out} = LO_2 - IF_{in} = RF_{in} + f_D$ and is again amplified with an RF amplifier. The resulting lower sideband is the desired Doppler shifted signal while the upper sideband is beyond the range of the USRP and is thus neglected. The generated velocity is $v_{gen} = f_D \lambda_c/2$. The antennas are spaced about 50 cm apart (less than the angular resolution), to make them appear as one target in the DOA image. RF absorbers are put between the antennas to avoid direct coupling.

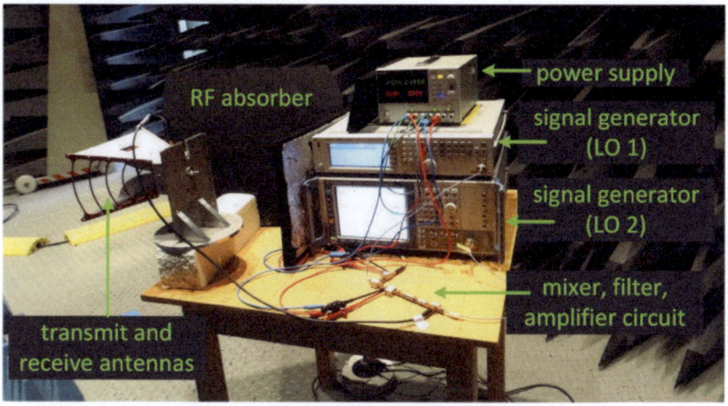

(a) Doppler emulator setup.

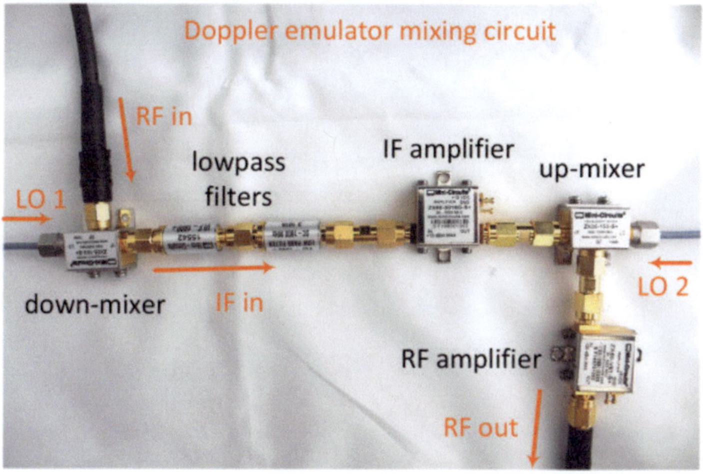

(b) Doppler emulator circuit. Power supply to amplifiers not shown.

Fig. 7.19: Chamber: Setup of Doppler emulator with signal generators, a mixer circuit and double-ridged horn antennas. The Radar signal is down- then up-converted with a slightly different LO frequency to produce the Doppler effect.

As shown in Fig. 7.20, the Doppler emulator setup is put at a distance of 3.7 m away from the radar while Target A and Target B are put 9.8 m and 6.3 m away respectively as markers. A radar estimation is first performed without activating the Doppler emulator, akin to an initial scenario of static objects or 'empty room' reference measurement. Fig. 7.21(a) and Fig. 7.21(b) show the radar image and DOA image of the empty room. In Fig. 7.21(b), the two antennas of the Doppler emulator are reflected at the correct distance of 3.7 m, with the other markers also at the correct distances and positions.

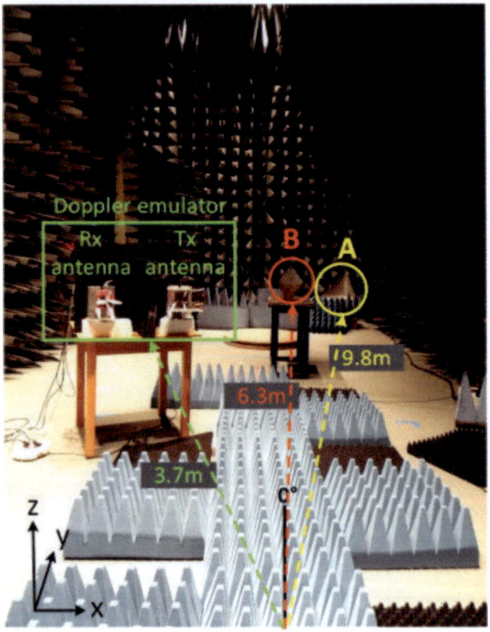

Fig. 7.20: Chamber: Azimuth DOA with velocity measurement setup with Target A and Target B as markers. The Doppler emulator setup is elaborated in Fig. 7.19.

The Doppler emulator is then activated to produce $v_{gen} = -50$ m/s to yield the results in Fig. 7.21(c) and Fig. 7.21(d). It is obvious in the radar image that a target with the stipulated velocity has been detected. Looking at Fig. 7.21(e), it is clear that the static targets and the reflection from the two antennas correspond to those of the empty room in Fig. 7.21(b). The Doppler emulator, now an active target is clearly seen at −50 m/s. Due to the proximity of its two transmit and receive antennas, which causes direct coupling, the active target appears as two side-by-side targets. One of the targets is located at the $\phi = -35°$ point and this in turn produced the ambiguity at the 35° point due to the high

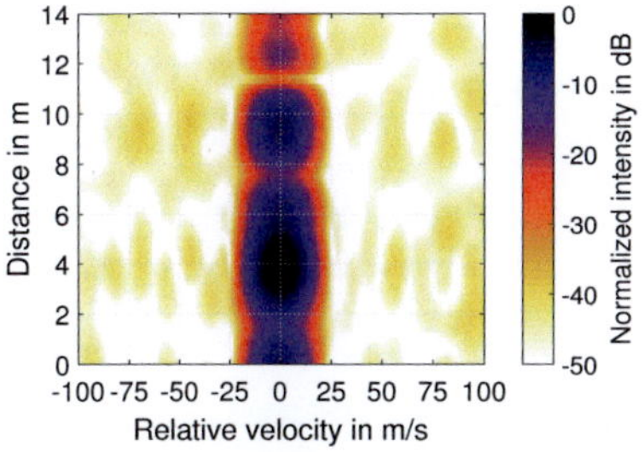

(a) Radar image: empty room.

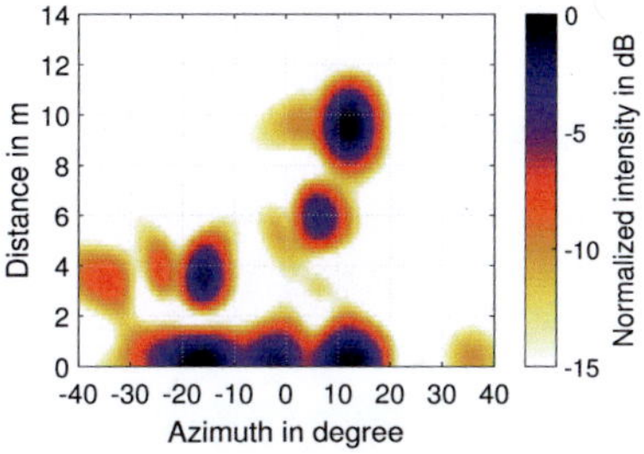

(b) DOA (slice at 0 m/s): empty room.

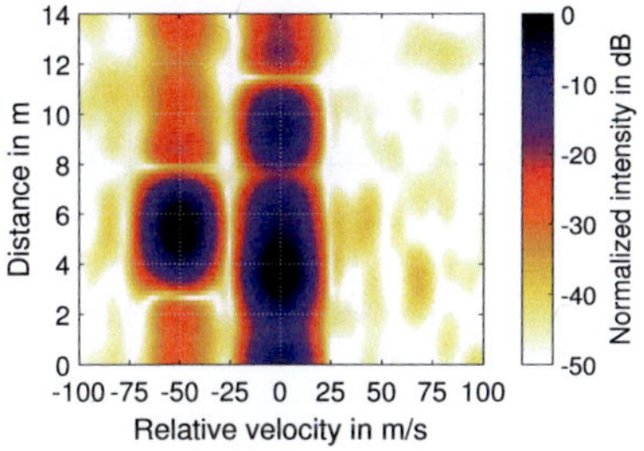

(c) Radar image: with Doppler.

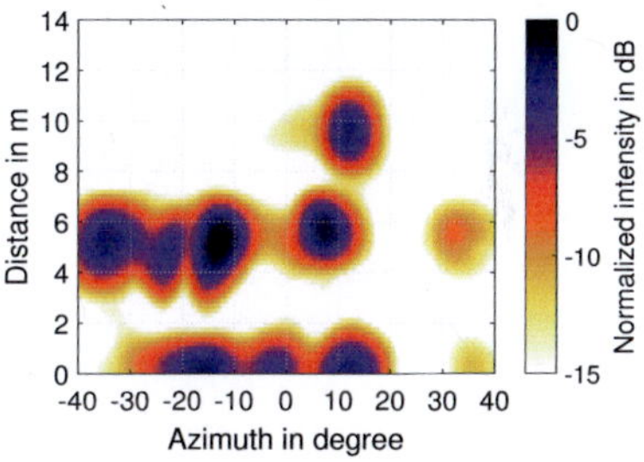

(d) DOA (summed): with Doppler.

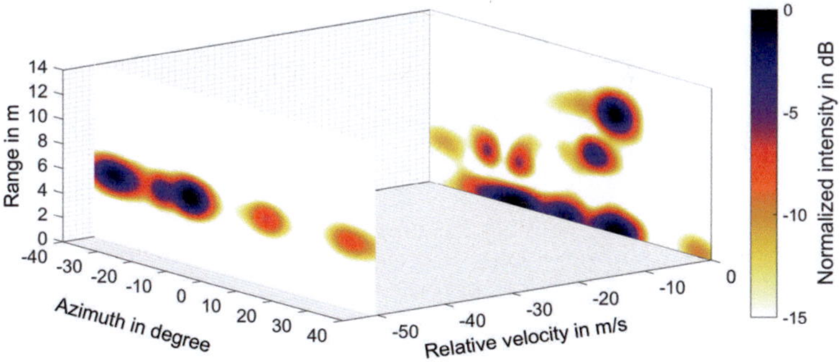

(e) DOA (slice): with Doppler.

Fig. 7.21: Chamber: 2D+velocity measurement according to setup in Fig. 7.20 for an empty room scenario (a) and (b), and for the case where the Doppler emulator is activated (c), (d) and (e). The DOA image slices in (e) show the static targets at 0 m/s while the Doppler emulator (active target) is at −50 m/s.

sidelobes. The resulting angle of the ambiguity can be cross-checked with the angular ambiguity plot of the patch antennas given in Fig. 7.8(a). These targets are also shifted by 2 m to 5.7 m due to Doppler emulator's circuitry delays.

This measurement has been done in only one transmit cycle, verifying that the 2D+velocity radar estimation is realizable with reliable estimation results.

7.5.4 Chamber: Interference cancellation measurement

This section aims to provide a measured result of the interference cancellation algorithm as well as a simulation verification of the measured results as a comparison. This measurement uses a real radar measurement from the EMC chamber with three targets as markers as shown in Fig. 7.22. The interferer signal is measured separately through the antennas and then added linearly to the signal from the real scenario in Matlab. The objective is to see how interference affects the radar outcome as well as to verify the performance of the interference cancellation algorithm.

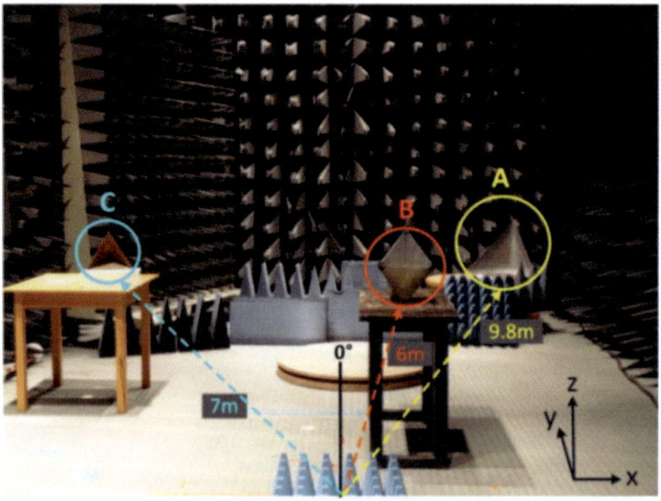

Fig. 7.22: Chamber: Azimuth DOA measurement setup for interference cancellation.

The Comm (interferer) is a four-transmitter node assigned with $u = \{4, 5, 6, 7\}$. The physical transmit array belonging to Comm is angled at 10° clockwise around the x-axis while looking directly at the Radar's receive array, which remains parallel to the x-axis. This adds a level of complexity in the signals since the propagation path length from each of the Comm's transmit antenna to each of the Radar's receive antenna will be different. The antenna arrays are spaced adequately apart to minimize the direct antenna coupling. By manipulating the gain blocks of the USRPs, the Comm's power is made to be of a certain level to achieve a maximum $\mathrm{SIR_{in}}$ of -19 dB. Due to the USRP imperfections, the Comm signal generated produced an $\mathrm{SIR_{in}}$ from the range of -14.5 dB to -19.3 dB. A $\delta_w T_0 = -0.3$ offset is also embedded into the Comm's signal. This signal is not equalized by its hardware's $\mathbf{H}_{\mathrm{cal}}$ to emulate a real scenario where this calibration matrix of a communication partner is not known to the Radar. It must be stressed here that the Comm signal is generated within the linear operation region of the USRPs.

The Comm signal that is received at the Radar's four Rx are then added linearly to the respective Rx in Matlab. This process is no different from having the Comm transmit at the same time as the Radar and has been tested and confirmed in [SNB$^+$16]. The main reason the Comm signal is generated and added separately from the Radar's reflected signal is due to the added degrees of freedom in observing and manipulating the Comm signal, something that is not possible with a simultaneous Radar and Comm transmission.

Fig. 7.23(a) and Fig. 7.23(b) show the radar image and the DOA image of the empty room (no Comm) respectively. Judging from all 16 radar images, the most stable $\mathrm{SNIR_{out}}$ values are from Tx 2 and Tx 3 of Rx 2, which are the best antennas shielded from the near-range clutter e.g. the metal tripod and RF absorbers. Hence the $\mathrm{SNIR_{out}}$ of the radar image taken at Rx 2 from Tx 3 will be taken as the reference. With no interference, the $\mathrm{SNIR_{out}}$ of the aforementioned radar image is $\mathrm{SNIR_{out,2,3}} = 54$ dB.

After the addition of the Comm signals, Fig. 7.23(c) and Fig. 7.23(d) result. It can be seen that the Comm signal with a frequency offset adds noise to the radar image and consequently distorts the estimation of the DOA image. The achievable $\mathrm{SNIR_{out}}$ with interference is now $\mathrm{SNIR_{out,2,3}} = 24$ dB.

The interference cancellation procedure described in Chapter 5 is then applied. Here, since Comm is a 4-Tx node, it can be assumed that all Tx will be shifted by the same amount of frequency offset. Hence each Radar Rx only need to estimate the XCORR and SCA once, then proceed to estimate the range and frequency residues using the channel coefficient matrices estimated from $u =$

$\{4,5,6,7\}$ in a loop. The estimated $\widehat{\mathbf{H}}_w^{\text{frame}}(n,m)$ differs slightly for each of the Radar's Rx and Comm's user index since the path traveled by the signals to the antennas are slightly different.

Finally, with 16 reconstructed Comm signals, each Rx will subtract four Comm signals and then put through the usual radar processing to yield Fig. 7.23(e) and Fig. 7.23(f). The interference canceled images have a lower SNIR_{out} than the empty room image at $\text{SNIR}_{\text{out},2,3} = 46\,\text{dB}$, an improvement of 22 dB from the image with interference. The SNIR_{out} loss is due to the near-field clutter, which elevates the noise floor, the absence of the calibration matrix from the Comm and also the large frequency offset, which distorts the $\widehat{\mathbf{H}}_w^{\text{frame}}(n,m)$ estimate. It is worth mentioning here that the BER is 0 for the demodulation of the Comm's data, signifying that a simultaneous radar estimation and communication is entirely possible with this system.

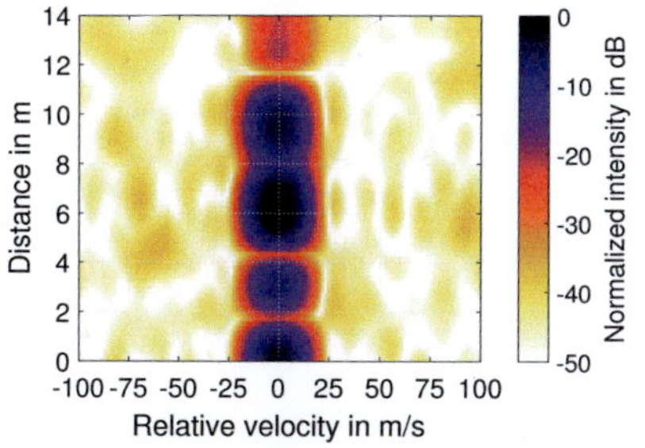

(a) Radar image: without interference, $\mathrm{SNIR}_{\mathrm{out},2,3} = 54\,\mathrm{dB}$.

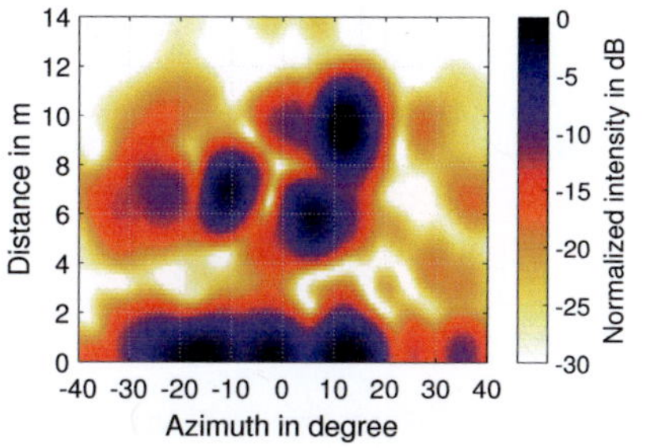

(b) DOA (slice): without interference

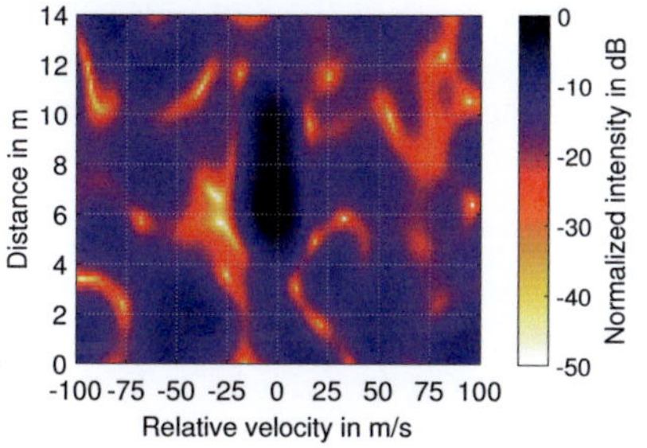

(c) Radar image: with interference, $\mathrm{SNIR}_{\mathrm{out},2,3} = 24\,\mathrm{dB}$.

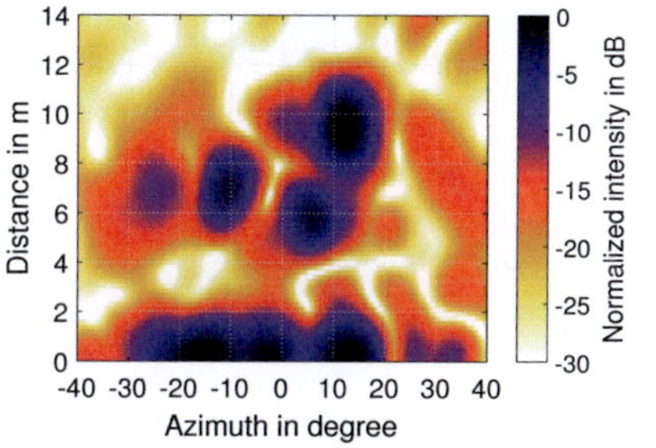

(d) DOA (slice): with interference.

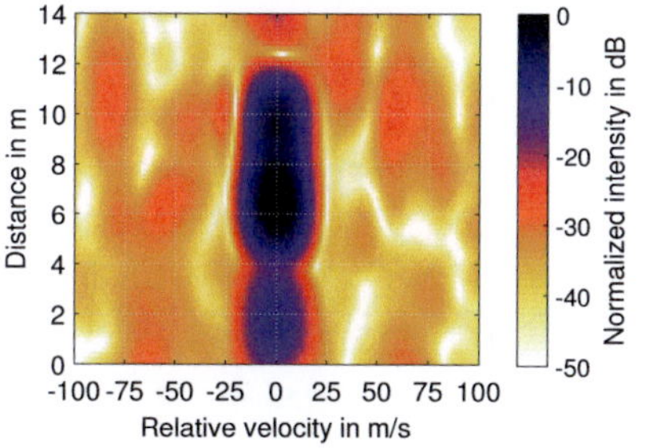

(e) Radar image: after interference cancellation, $\mathrm{SNIR}_{\mathrm{out},2,3} = 46\,\mathrm{dB}$.

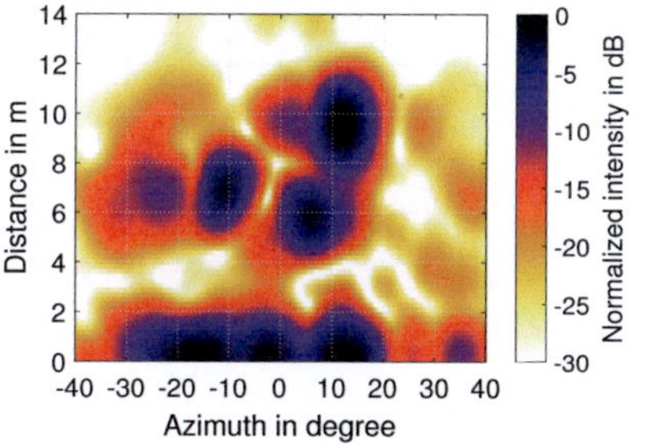

(f) DOA (slice): after interference cancellation.

Fig. 7.23: Chamber: Measured interference cancellation from another 4×4 MIMO RadCom node with $\delta_w T_0 = -0.3$ and $\mathrm{SIR}_{\mathrm{in}} = [-19.3, -14.5]\,\mathrm{dB}$. $\mathrm{SNIR}_{\mathrm{out},2,3}$ refers to the $\mathrm{SNIR}_{\mathrm{out}}$ taken at Rx 2 from Tx 3. All DOA images are slices from 0 m/s.

Simulation verification of interference cancellation measurement

A simulation of the measurement scenario in Fig. 7.22 was set up in Matlab to verify the result of the interference cancellation. To emulate the wireless measurement scenario as closely as possible, certain effects as follows are taken into account:

- referring to Fig. 7.23(a), the $\mathrm{SNIR}_{\mathrm{out},2,3} = 54\,\mathrm{dB}$ of the Radar without any interference encompasses the Radar's received signal along with the antenna gains, amplifier gains, system noise figure, processing gain and windowing loss. Hence the actual power level of the signal minus the processing gain $G_{\mathrm{per}} = 45.15\,\mathrm{dB}$ and the Hamming windowing loss of 2.7 dB, is approximately 6 dB, which is then the SNR at the Rx set in the Matlab simulation.
- Target A, B and C are modeled as point targets and with their RCS approximated via a heuristic approach to produce the $\mathrm{SNIR}_{\mathrm{out},2,3} = 54\,\mathrm{dB}$. This yields for Target A, B and C respectively, $11.5\,\mathrm{dBm}^2$, $-1.5\,\mathrm{dBm}^2$, and $0.8\,\mathrm{dBm}^2$. Compared to their respective maximum values in Table 7.2, this is a significant decrease but is entirely feasible due to their orientations to the plane of incidence of the Radar. Target A is for example, placed lower than the antennas's height, making its orientation around $-10°$ to $-30°$ away from its axis of symmetry. Using the measurement charts shown in [Rob47], taking the angle of orientation as $-20°$ and azimuth angle $\phi = 12°$, the chart shows a loss of around 5 dB in power. Since Target A is also blocked by some RF absorbers, a further reduction is to be expected. Target B has a more complicated response that is highly dependent on its orientation. While Target C suffers the least RCS loss because it has been tilted toward its axis of symmetry to the Radar.
- the reflections from the near-range clutter as stipulated in Table 7.3 are taken into account.
- the antenna coupling between the transmit antennas and the transmit to receive antennas, with the values as given at the beginning of this section are taken into account.
- the Comm signal undergoes a two-path propagation model. Considering the setup of the Comm's transmit array and the Radar's receive array in terms of their distance and height from the ground, the reflected path

from the ground is found to be around two times the line-of-sight (LOS) distance from the Tx of the Comm to the Rx of the Radar.

- the Comm signal is set to produce an exact $\mathrm{SIR_{in}} = -19.3\,\mathrm{dB}$ with $\delta_w T_0 = -0.3$.

The simulation setup produced the results as shown in Fig. 7.24. Comparing the $\mathrm{SNIR_{out}}$ of the radar images of the measured and simulated results in Table 7.5, it can be seen that there is a 3 dB difference in the radar image with interference and 1 dB difference after interference cancellation. This can be attributed to several factors such as:

- USRP imperfections e.g. the aliasing of the transmit signal of the USRP. Operating at 100 Msps to achieve 100 MHz analog bandwidth, the imperfect filter roll-off of the USRP contributes to the aliasing of both the Radars and Comm's transmit signals.
- imperfect modeling of the near-range clutter.
- USRP imperfections in the gain blocks and different cable losses contribute to the $\mathrm{SIR_{in}}$ difference at each Rx port of the USRP.
- multipath propagation not taken into account for the Radar's signal and not sufficiently modeled for the Comm signals.
- secondary reflections from the targets to the metal floor or surroundings not taken into account.

The above factors cause an increase of the noise floor for both the Radar and the Comm's signals. It was found in the simulation that the non-line-of-sight (NLOS) reflected path of the Comm signal easily contributed 1 dB to the noise. It can be seen that the DOA images with interference are significantly different from their original or interference canceled counterparts, and is very depending on the signal paths contribution from Comm. Hence a comparison cannot be made in this case for Fig. 7.24(d) with Fig. 7.23(d).

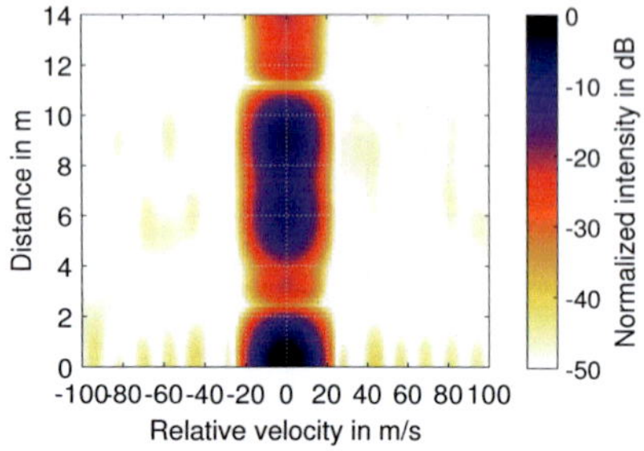

(a) Radar image: without interference, $\mathrm{SNIR}_{\mathrm{out},2,3} = 54\,\mathrm{dB}$.

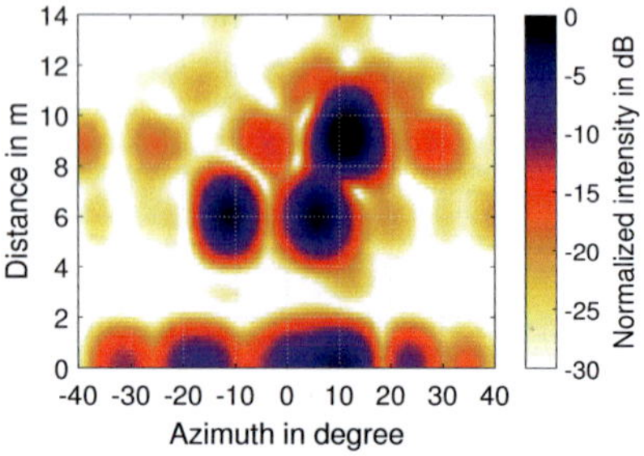

(b) DOA (slice): without interference

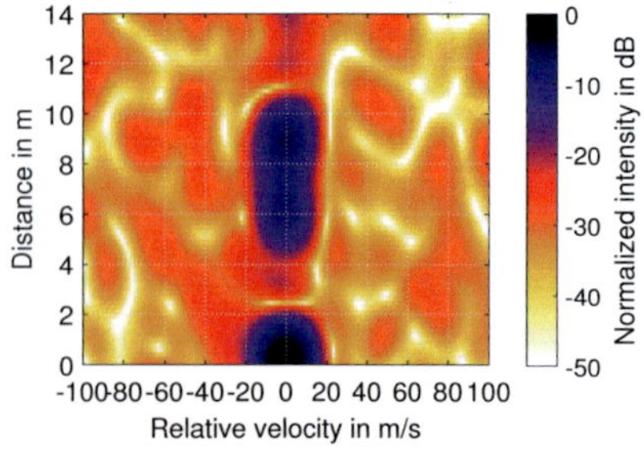

(c) Radar image: with interference, $\mathrm{SNIR}_{\mathrm{out},2,3} = 27\,\mathrm{dB}$.

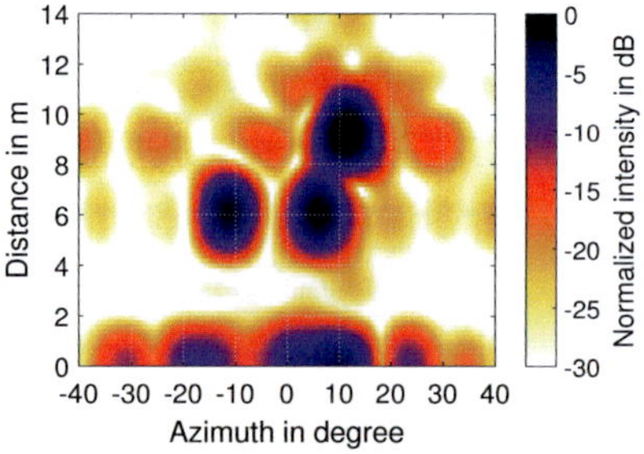

(d) DOA (slice): with interference.

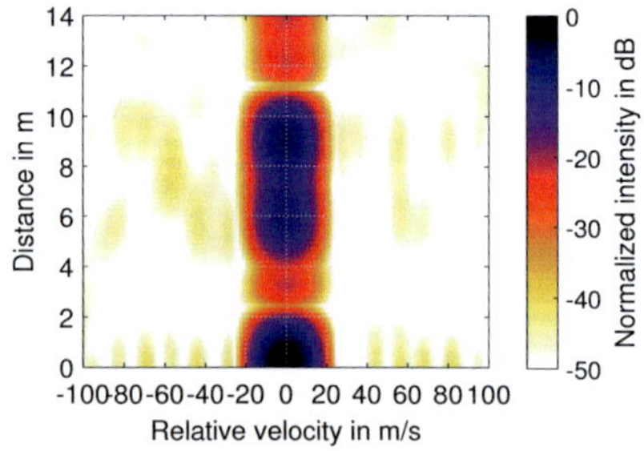

(e) Radar image: after interference cancellation, $\mathrm{SNIR}_{\mathrm{out},2,3} = 47\,\mathrm{dB}$.

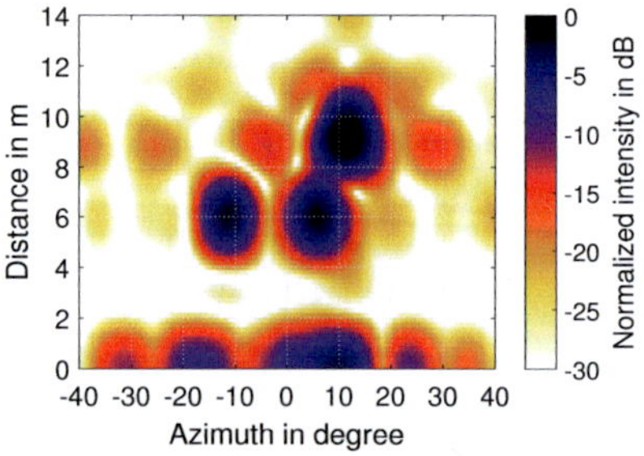

(f) DOA (slice): after interference cancellation.

Fig. 7.24: Chamber: Simulation verification of the interference cancellation in Fig. 7.23 from another 4×4 MIMO RadCom node with $\delta_w T_0 = -0.3$ and $\mathrm{SIR}_{\mathrm{in}} = -19.3\,\mathrm{dB}$. Taken into account for the simulation was a two-path propagation model for Comm, transmit-to-receive antenna coupling for the Radar, and transmit-to-transmit antenna coupling for the Radar and Comm and SNR = 6 dB. All DOA images are slices from 0 m/s.

Comparison between measured and simulated interference cancellation results

The interference canceled DOA results however appear similar with a difference of only 1 dB. A comparison between the measured and simulated target properties is summarized in Table 7.5 and Table 7.6.

Table 7.5 shows the normalized target peaks. In the simulated results, interference cancellation can cause the loss of amplitude in certain point targets. The same cannot be said for the measured results since the targets consist of a cluster of reflection points, hence their losses tend to be larger and less uniform.

Looking at Table 7.6 of the simulation results, it can be seen that the interference cancellation algorithm does not change the properties of the targets' positions. In the measured results however, a slight change in estimated angle $\hat{\phi}$ and r can be seen for Target A and C. This again, is due to the multiple reflections from the target (due to the different reflection centers) to the Radar. Since the amplitude loss is not uniform over all reflections coming from the same target, certain rays tend to experience less losses and hence the maxima of the target in the DOA image also shifts slightly. The slight differences are confined within the diffuse points of the targets, hence they pose no problem toward the accuracy of the radar estimate.

Table 7.5: Chamber: Comparison of radar image $SNIR_{out}$ and normalized target peaks between simulated and measured interference canceled results.

Parameter	Measured			Simulated		
	Original in dB	Interfered in dB	Canceled in dB	Original in dB	Interfered in dB	Canceled in dB
$SNIR_{out}$	54	24	46	54	27	47
peak A	0	-	-0.5	0	-	0
peak B	-2	-	-2.2	-2.2	-	-2.2
peak C	-2.3	-	-1.2	-2.4	-	-2.7

Table 7.6: Chamber: Comparison of target positions (estimated angle $\hat{\phi}$ and range r) between the original and interference canceled DOA results.

Target	Measured				Simulated			
	Original		Canceled		Original		Canceled	
	$\hat{\phi}$ in °	r in m	$\hat{\phi}$ in °	r in m	$\hat{\phi}$ in °	r in m	$\hat{\phi}$ in °	r in m
A	11.5	9.6	12	9.6	11.5	9.2	11.5	9.2
B	6	5.8	6	5.8	6	6	6	6
C	-11	6.9	10	7.3	-11	6	-11	6

Tolerance chart for a measured MIMO Radar with a 4-Tx Comm interferer

By varying the Comm's SIR_in and frequency offset, a tolerance chart for the Radar occupying $u = \{0,1,2,3\}$ and Comm $u = \{4,5,6,7\}$ is plotted and shown in Fig. 7.25. With interference from a MIMO node, the tolerance of the Radar toward frequency offset becomes even lower. An offset of around $\delta_w T_0 = [-0.05, 0.05]$ irrespective of the SIR_in will give a reasonable SNIR_out. Apart from the small window of tolerance the interference cancellation algorithm needs to be applied to obtain a reliable estimate.

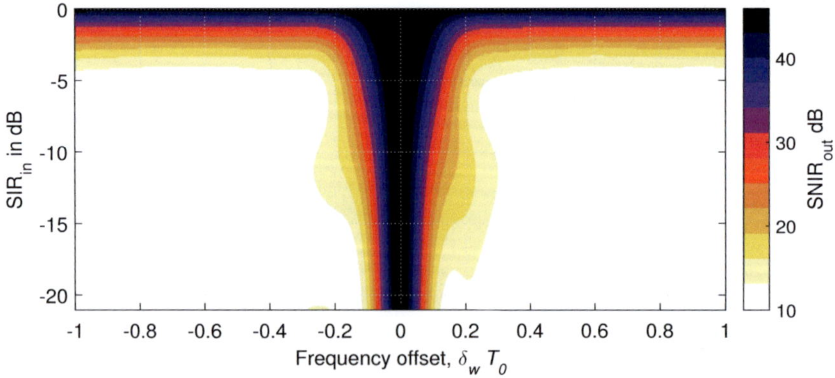

Fig. 7.25: Measured tolerance plot: SNIR_out vs. SIR_in for 4-Tx MIMO Comm at $-1 \leq \delta_w T_0 \leq +1$ without interference cancellation. Each level of color represents 3 dB.

7.6 Outdoor 1: DOA measurement of static targets

In the same way as the measurements in the chamber are done, the MIMO RadCom setup on USRP is now moved outdoors. The first scenario deals with the measurement of static reflectors as targets. Within the same scenario, Target A is moved further to the right (thr real azimuth angle ϕ becomes larger) as shown in Fig. 7.26 while Target B and C remain at their initial positions. Target A is initially at 12.5 m while Target B and C are 6 m and 3.5 m away from the Radar respectively.

Fig. 7.27(a) shows the empty room DOA measurement, normalized by the highest point of the subsequent measurements with targets (i.e. from Scene 1b). For the Scene 1a, Fig. 7.27(b) results. Target A is estimated to be located at $\hat{\phi} = 12°$ but due to its orientation toward the Radar, some secondary reflections (or reflections from ground clutter) at $\phi > 30°$ results and this leads to the ambiguity at $\hat{\phi} = -67°$ caused by the sidelobes (refer to the antenna arrays' angular ambiguity plot Fig. 7.8(a)).

In Scene 1b, Target A is shifted to $\hat{\phi} = 19.5°$ to yield Fig. 7.27(c). Again, there are weaker secondary reflections, which cause the ambiguity at $\hat{\phi} = -68°$. Some minor reflecting objects were also in the vicinity of Target A and causes the ambiguity at $\hat{\phi} > -80°$.

In the final Scene 1c, Target A is shifted to $\hat{\phi} = 40°$, clearly out of the Radar's angular FOV. As expected, a corresponding ambiguity emerged at $\hat{\phi} = -57°$ in Fig. 7.27(d). All ambiguities that result from the reflections or secondary reflections of Target A that are $\phi > 30°$ are consistent with the locations given by Fig. 7.8(a).

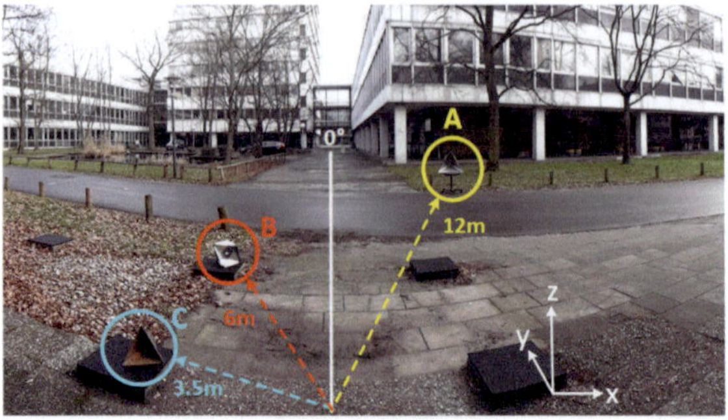

(a) Scene 1a.

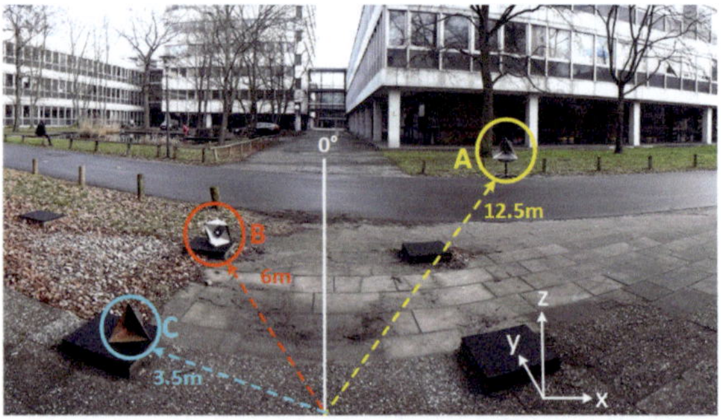

(b) Scene 1b.

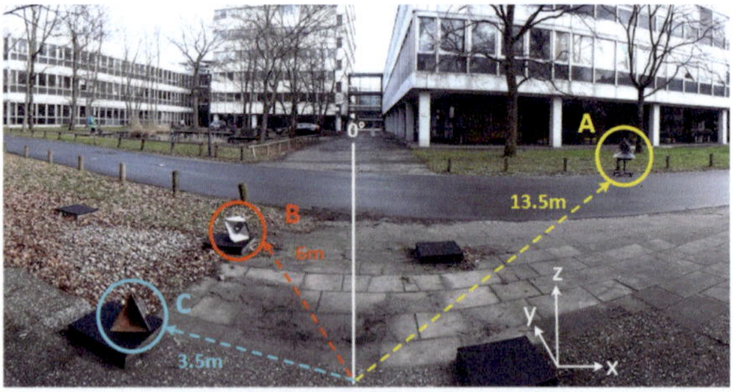

(c) Scene 1c.

Fig. 7.26: Outdoor 1: MIMO DOA measurement of static targets. Target A is shifted farther down the x-axis direction to the point that it is out of the unambiguous angular field-of-view of the Radar.

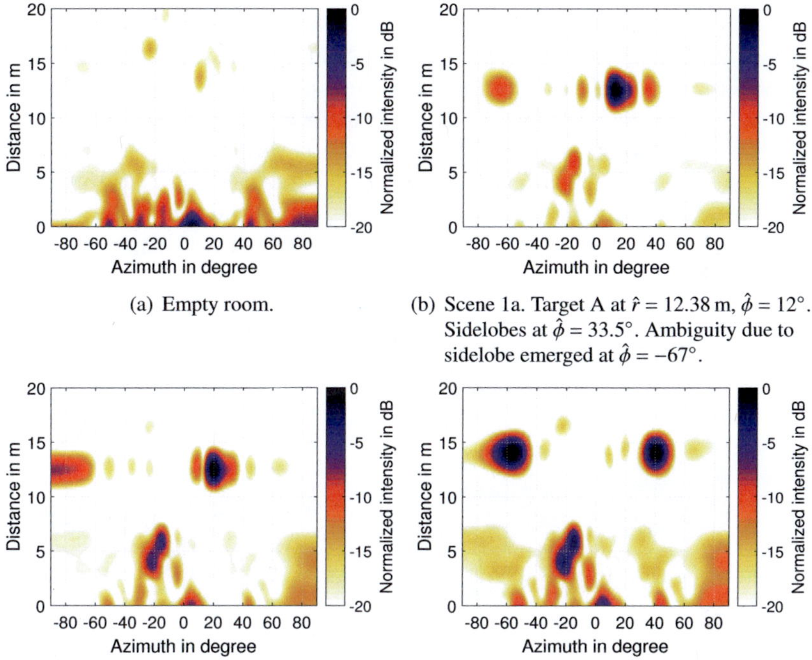

(a) Empty room.

(b) Scene 1a. Target A at $\hat{r} = 12.38$ m, $\hat{\phi} = 12°$. Sidelobes at $\hat{\phi} = 33.5°$. Ambiguity due to sidelobe emerged at $\hat{\phi} = -67°$.

(c) Scene 1b. Target A at $\hat{r} = 12.56$ m, $\hat{\phi} = 19.5°$. Sidelobes at $\hat{\phi} = 33.5°$ and 42.5° Ambiguity due to sidelobes at $\hat{\phi} = -68°$ to $-80°$.

(d) Scene 1c. Target A at $\hat{r} = 13.8$ m, $\hat{\phi} = 40°$ (exceeding antenna's angular unambiguous region). Ambiguity at $\hat{\phi} = -57°$.

Fig. 7.27: Outdoor 1: Measured DOA image slices at 0 m/s from the estimated angles of $\hat{\phi} = -90°$ to 90°. All normalized to the DOA image of Scene 1b for comparison. The shifting of Target A farther down the x-axis and out of the unambiguous angular field-of-view results in a distinct ambiguity in (d).

7.7 Outdoor 2: DOA measurement of a moving target with interference cancellation

The final outdoor measurements deal with a moving target and a 1-Tx Comm as the interferer. Fig. 7.28 shows the scenario where a van passes through the Radar from behind, traveling at a speed of 25 to 35 km/h. In order to increase the velocity resolution for a better discernment in the radar image, the OFDM symbol length is increased to $4M$ to result in $\Delta v = 1.57$ m/s. The radar image captured is as shown in Fig. 7.29(a) with the corresponding summed DOA image of the moving and non-moving targets shown in Fig. 7.29(b). Based on the location of the target with velocity in the radar image, it can be seen that the moving target is located at around $\phi = -10°$ of $r = 45$ m in the DOA image. Other 'targets' in the image arise from the reflections of various metal objects in the scenario as pointed out in Fig. 7.28. Fig. 7.30(c) clearly shows the locations of the above mentioned moving and static targets, using DOA image slices from different velocities. The maximum SNR_{out} is 48 dB.

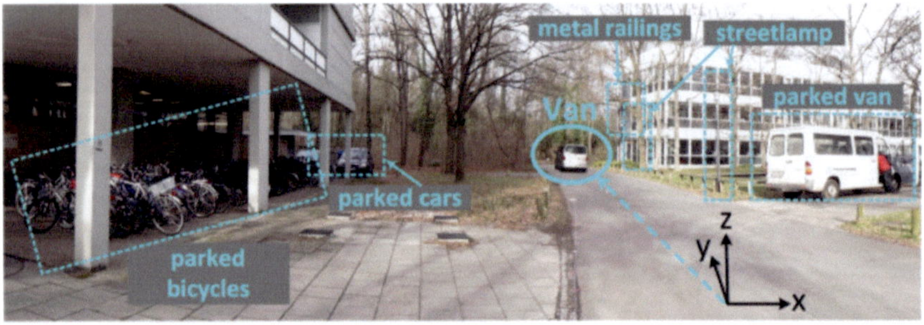

Fig. 7.28: Outdoor 2: Outdoor measurement scenario of the moving target (van) that is coming from behind the Radar with various static reflectors (parked vehicles, street lamps and signs).

Using the same method as mentioned in Section 7.5.4, a Comm signal at $u = 4$ with the offset of $\delta_w T_0 = -0.3$ is manipulated to give an SIR_{in} of −21 dB. With the Comm signal added to the Radar's reflected signal, the radar images are as shown in Fig. 7.30(a) and Fig. 7.30(c). The maximum SNR_{out} is 31 dB, consistent with the reading from the tolerance chart given in Fig. 7.2.

The interference cancellation algorithm is run for each Rx at the Radar's side and the corresponding reconstructed Comm signal is subtracted. They yield Fig. 7.30(b) and Fig. 7.30(d). The maximum SNR_{out} is now 47 dB, an increase of 16 dB. In the same way as demonstrated in Section 7.5.4 for static targets, the interference cancellation algorithm has been demonstrated to work even in a moving target scenario.

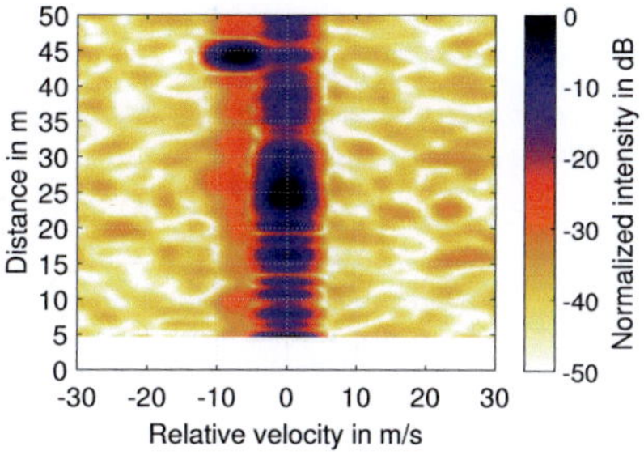

(a) Radar: original. $SNR_{out} = 48$ dB.

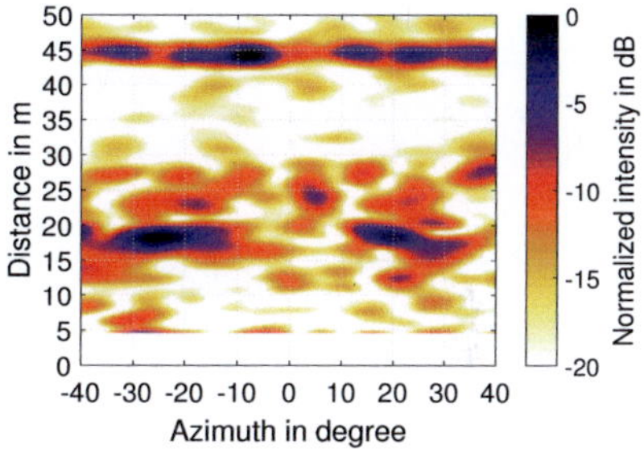

(b) DOA (summed): original.

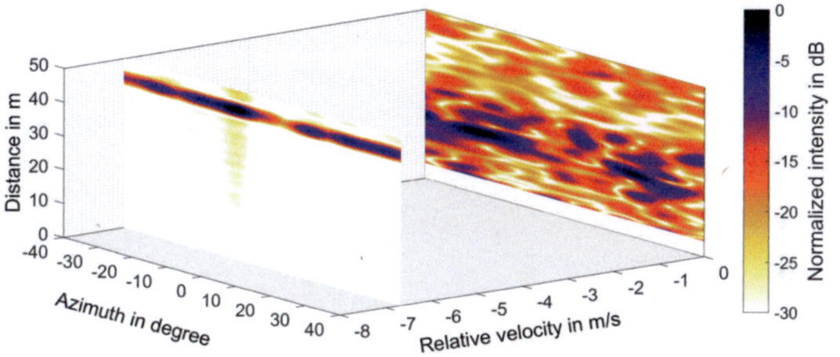

(c) DOA slices: original. Moving target is at −6.9 m/s, static targets and clutters are at 0 m/s.

Fig. 7.29: Outdoor 2: Measured radar image and DOA image of a moving target (van) without interference. The results from 0 to 5 m were removed due to strong reflection from the ground and RF absorbers directly in front of the antenna arrays.

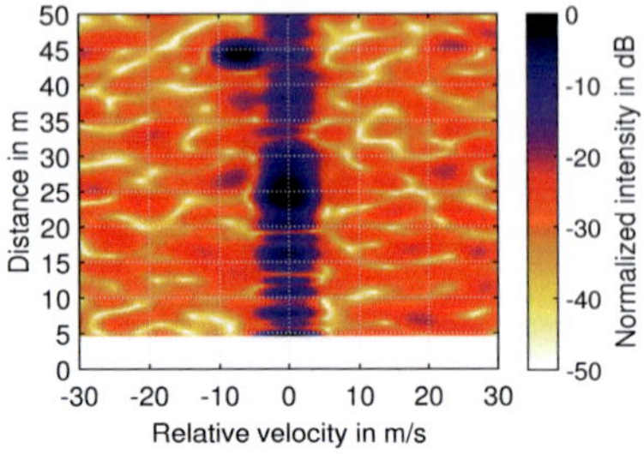

(a) Radar: with interference.
$SNR_{out} = 31$ dB.

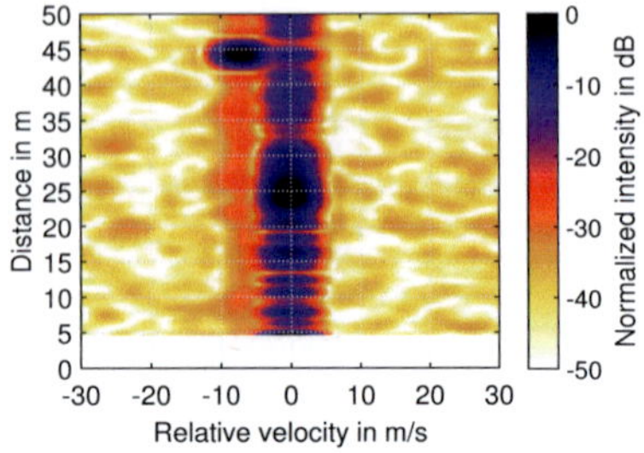

(b) Radar: interference canceled.
$SNR_{out} = 47$ dB.

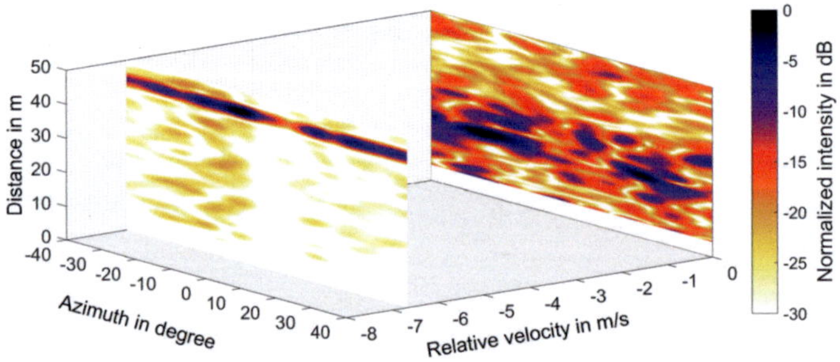

(c) DOA image (slice) with interference.

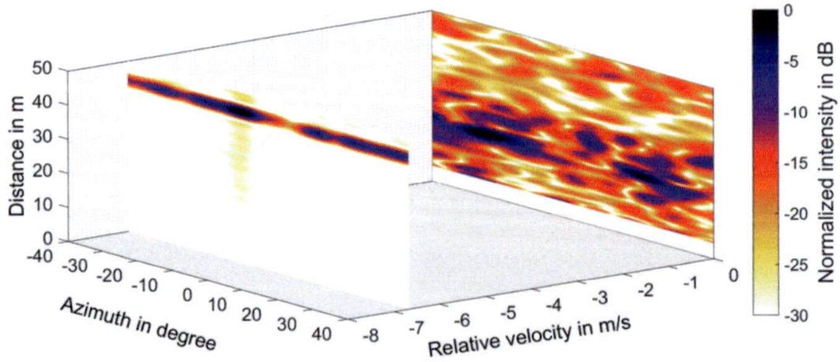

(d) DOA image (slice) after interference cancellation.

Fig. 7.30: Outdoor 2: Measured radar and DOA images of a moving van before and after interference cancellation with 1-Tx Comm (interferer) with $\delta_w T_0 = -0.3$ that results in $SIR_{in} = -21$ dB. The van is shown at -6.9 m/s of the DOA image slice, while all static objects and clutter are at 0 m/s.

7.8 Chapter 7 summary

Measurement verifications of the MIMO RadCom system using a real-time 100 MHz instantaneous bandwidth setup on USRPs have been presented in this final chapter. At the point of writing, these measurement verifications with static and moving targets are the first reported in literature. Due to the real-time simultaneous transmit and receive setup, these measurement results also take into account the effect of the physical antennas, signal model, user index, hardware and channel imperfections, something that is impossible with the time-division multiplexed technique currently used by MIMO radar researchers.

The signal processing is not automated together with the real-time measurements but is always done on site immediately after every measurement, requiring about 3 s for a 2D+velocity processing and 4 s per Comm signal loop in the interference cancellation. The only reason the signal processing step is not automated is because the super computer is overclocked to synchronize four USRPs at the maximum achievable full duplex bandwidth thus it cannot run another application in parallel.

To substantiate the interference cancellation algorithm, wired measurements were made to corroborate the interference tolerance chart and the Case Studies, which involved three different setups. It has been proven through these measurements that the interference cancellation algorithm functions as expected even with signals generated by imperfect hardware. The imperfection distorts the signals non-linearly yet the interference cancellation algorithm is able to separate the distorted Comm (interferer) signal adequately to improve the radar $SNIR_{out}$, albeit with a slightly reduced performance compared to the simulated results.

In a controlled environment i.e. in the EMC chamber, wireless measurements relating to angular resolution, DOA in both the azimuth and elevation planes, velocity and interference cancellation are demonstrated. Only the 2D+velocity was demonstrated since a 4×4 3D measurement will not provide enough angular resolution. The interference cancellation measurement has also been substantiated with a simulation verification, which again verified the capability and capacity of the proposed interference cancellation algorithm even in a wireless environment, where the non-linear distortions to the signal are even more severe. With the advantage of a real-time system, the breakthrough of this work are the 2D+velocity measurements were also performed outdoors with static and moving targets. In this outdoor environment with a moving target, the interference cancellation was also successfully applied.

8 Conclusions

To conclude this thesis, the major contributions and findings according to the motivation and goals presented in the Introduction chapter will be recapitulated.

8.1 Contributions

Signal model

In Chapter 2 a detailed analysis of the signal model for use in a multi-user and multi-antenna node was derived and simplified. These were not considered in [Stu12, Bra14] since the authors did not consider a MIMO radar configuration hence these additional effects will not emerge in the SISO radar estimation. It was found that:

- the range of a target causes a phase that rotates only along the frequency-subcarrier axis of the OFDM frame.
- the Doppler of a moving target incur a significant phase rotation over the time-symbol axis and a frequency offset at the frequency-subcarrier axis.
- due to the proximity of the collocated antenna array, each of the qp-th receive frame is expected to show the same radar (range-Doppler) image as long as there is no other environmental factors such as transmit-to-receive antenna coupling.
- the position of the target merely adds a fixed phase offset to the whole frame, slightly differing for each qp-th frame due to the minuscule difference in the path lengths and this is used for the DOA estimation.

- the user index u, that is used to define the subcarrier allocation to all transmit antennas in Chapter 2 introduces an extra constant phase shift over the received qp-th frame and this in turn affects the accuracy of the angular estimation as shown in Chapter 6.
- the incorrect removal of the CP length affects the accuracy of the range estimation. For every sample of CP that remains prepended to the received signal-under-evaluation, the range error will increase by one range resolution bin.

With the OFDM scheme being very susceptible to subcarrier misalignment (arising from unavoidable frequency offsets from hardware imperfection, environment and moving targets/nodes), it is foreseeable that an interference cancellation algorithm will definitely be required to improve the fidelity of the radar signal, just as required by its communication system counterpart. All the aforementioned radar channel effects have been analyzed to determine their co-dependency with each other so that a suitable interference cancellation algorithm can be designed.

Influence of physical antennas

It is inevitable that a DOA estimation is also dependent on the radiation pattern characteristics produced by the physical antennas and therefore must be expanded. The placement of the transmit and receive arrays to form the simplest case of no-redundancy virtual antennas using uniform linear arrays was presented for both the cases of a 1D (for either azimuth or elevation detection only) and 2D (for both azimuth and elevation) antenna configurations.

From the physical antenna geometry, a beamsteering vector that equalizes the phases due to the antenna element spacing was also derived. The purpose of the derivation is to provide the fundamentals to the virtual antenna concept and hence the resulting beamsteering vector since this information is not widely available or known. Hence Chapter 3 focused on the systematic derivation of expressions, which can be adapted to optimize the antenna arrays, and showed the simplified forms of equations whenever possible for a direct implementation in a signal processing software i.e. Matlab.

It has also been proven through simulations that:

- 1D configuration featuring 16 virtual antennas is the minimum number to obtain a reasonable DOA estimate.
- with the ideal spacing of $d_\mathrm{R} = \lambda_\mathrm{c}/2$ with $d_\mathrm{T} = Qd_\mathrm{R}$, the unambiguous region was shown to be around $\phi, \theta = [-70°, 70°]$ for a 1D uniform linear array (ULA) of 4×4 configuration, obtained through the angular ambiguity plot.
- when the ambiguity plot is adjusted with real antenna array parameters in the measurement verification, the unambiguous angular regions were found to be the same as the theoretical ones.
- since the minimum elements per angular plane is 16, the 2D array with an antenna count per angular plane that is lower than this number will suffer worse angular resolution and ambiguity regions. For this reason, only a 2D+velocity (no elevation) real-time measurement verification was performed since only up to four USRPs can be supported by the host computer at this point of time.

Tolerance toward communication partners

A brief analysis of the communication channel using the expressions already derived for the radar channel in Chapter 4 showed that with a little frequency offset, the output signal-to-interference and noise ratio ($\mathrm{SNIR_{out}}$) of the radar will decrease. The amount however depends on the input signal-to-interference ratio ($\mathrm{SIR_{in}}$) of the radar signal, something which was not analyzed in [SSBZ13]. Since a loss of subcarrier orthogonality is highly probable in a real scenario, a tolerance chart (Fig. 4.7) featuring the change of $\mathrm{SNIR_{out}}$ over $\mathrm{SIR_{in}}$ and frequency offset due to the Comm signal was created to give an insight to the robustness of the Radar. This chart was simulated using the worst case scenario of the spectrally interleaved signal model - when the Radar and Comm are allocated directly adjacent subchannels. From the communication channel analysis and the tolerance chart it can be deduced that:

- the Radar's $\mathrm{SNIR_{out}}$ is dependent on the amount of power reflected from the targets (i.e. the target's RCS). Assuming the Radar and Comm transmit the same power, the $\mathrm{SIR_{in}}$ is proportional to square of the distance of the Comm from the Radar.

- the Comm signal at the Radar's receiver will in most cases always be higher than the desired backscattered signal.
- the tolerance chart shows that for a very small frequency offset of $\delta_w T_0 < 0.01$, the SNIR_{out} will stay above 30 dB no matter how strong the Comm signal is. This frequency offset corresponds to a Doppler shift caused by a target moving at 20.5 km/h (5.7 m/s) with the carrier frequency of 24 GHz. For larger frequency offsets, the influence of the power of the Comm signal becomes significant.
- training symbols and pilot tones are added to the OFDM frame to aid payload data recovery in the communication and radar systems.

Interference cancellation algorithm

Unlike for MIMO OFDM communication systems, where numerous interference cancellation algorithms have been proposed, to date, this work is the only research dealing with interference cancellation for MIMO OFDM radars, as mentioned in [BTJ13]. While radar networks are not yet the norm, in all eventualities, an interference cancellation algorithm must accompany this future radar concept due to the low tolerance of OFDM toward interference.

The idea behind the interference cancellation algorithm is to reconstruct the payload data with all its channel effects and then subtract it from the Radar's received signal. Based on the detailed analysis of the signal model in Chapter 2, the interference cancellation presented in Chapter 5 can be summarized to:

- detecting the start point of the Comm signal, which may be using either the same or different set of Schmidl and Cox algorithm (SCA) training symbols as the Radar. A cross-correlation set with a threshold will reveal peaks signifying the possible start points. Searching for only paired peaks (a condition imposed due to the nature of the cyclic prefix), all pairs are ranked according to their peak amplitude. The paired peak with the highest amplitude is chosen as the most probable start point.
- the frequency synchronization using the Schmidl and Cox Algorithm (SCA) is modified to take in the start time value from the previous step to prevent the algorithm from running into timing errors. The SCA becomes unreliable when the Comm signal power seen at the Radar's Rx is not sufficiently high.

- the residue phases due to the imperfect time and frequency synchronization are extracted with the help of the pilot tones and are corrected.
- a channel equalization matrix equalizes the leftover time and frequency offsets and signal amplitude for the signal frame and its training symbols.
- the payload data (along with training symbols and pilot tones) is recovered then reconstructed with all the estimated channel effects added in reverse and subtracted from the Radar's signal in time domain.
- case studies featuring the interference cancellation of a Comm with one- and multiple transmitters show that as long as the power of the Comm is 5 dB higher than the Radar's there is no error in the payload data demodulation (no channel coding used) and the $SNIR_{out}$ improvement is dependent on the severity of the frequency offset only.
- the proposed algorithm is of low-complexity and is capable for use in real-time processing with optimized functions and parallel processing.

Subcarrier allocation strategy

At the application level in Chapter 6, it is foreseen that a RadCom node might be supplied with more antennas than it is allowed to use. An 8×4 configuration was used as an example, where at any one time, only four transmitters can be active. Along with the choice of user index to be distributed to the node, the subcarrier allocation strategy showed that:

- the activated transmit antennas should be the ones closest to the origin of the array and are contiguous e.g. $p = \{2,3,4,5\}$ out of $p = 0, ..., 7$.
- the user index allocated to these transmit antennas should be contiguous e.g. $u = \{0,1,2,3\}$ or vice versa.
- the accuracy and angular resolution of the DOA is least affected when the above conditions are fulfilled but the DOA estimate error is still proportional to the target's distance r, and the system's subcarrier spacing Δf. At $r = 200$ m and $\Delta f = 90.9$ kHz, the error is 3.5°, which is still within the angular resolution.
- the small angle error due to the phase of the user index can be rendered insignificant with good parameterization such as the use of more transmit and receive antennas, and increasing the duration of the OFDM symbol

T_0. A compensation scheme in the post-processing can also be used to correct the phase offset due to the user index.

To wrap up the theoretical analysis, an example target tracking application utilizing the MIMO RadCom's 2D+velocity capability (simulation-based) was presented. The tracking algorithm is of low-complexity and the RadCom system is parameterized for real-time implementation on the ETTUS X310 USRP. When multiple nodes are networked together, the estimation is foreseen to have higher target localization accuracy with lower target obsurity through a 3D data fusion algorithm proposed in [NSZ16].

Measurement verification

Chapter 7 presents the measurement verifications of the MIMO RadCom system using a real-time 100 MHz bandwidth setup on USRPs. 'Real-time' refers to the simultaneous transmit and receive operations with four Tx and four Rx, thereby allowing all hardware and channel imperfections to be taken into account in the measurement verifications. To date, this is the first such reported real-time MIMO radar measurement with with static and moving targets in an outdoor environment. This breakthrough is significant as it paves a way for current MIMO radar researches to be verified through measurements using a commercially available yet affordable hardware. With this, a full-fledged working MIMO radar is now one step closer to reality.

To substantiate the MIMO RadCom concept that is capable of estimating range, Doppler and DOA within one transmit cycle, the following measurement verifications were preformed.

- Comparison of the measured and simulated RadCom interference tolerance chart showed that they are almost the same and thus can be used as a guide irrespective of the frequency, bandwidth and subcarrier spacing.
- Comparison between the simulated and measured interference cancellation algorithm based on the three Case Studies presented in Chapter 5 confirmed that the algorithm also works well with real, non-linearly distorted signals.
- Two patch antenna arrays for transmit and receive were analyzed using the tools presented in Chapter 3 to verify its theoretical angular resolution ($\approx 5.3°$) and unambiguous operating angles ($\phi, \theta = [-30°, 30°]$).

- DOA simulations show that near-range clutter (either from specular or diffuse reflections) at $r = [0,2]$ m from the Radar tend to emerge with multiple lobes on the DOA image. This takes the diffuse reflections from the ground and metal tripods with the antennas into account in the noise floor and SNIR_{out} of the radar and DOA images.
- Direct transmit to receive antenna coupling with RF absorbers put in between the arrays ranges from -83 to -73 dB as measured with a vector network analyzer and is also taken into account.
- Indoor measurements show that the angular resolution is approximately 8.5°, which is a realistic value. All other indoor measurements with static targets (corner reflectors) and Doppler emulator confirmed that the MIMO RadCom can indeed perform real-time 2D+velocity measurements with sufficient accuracy.
- An interference cancellation measurement was also performed and verified by a simple simulation in Matlab that takes all the above mentioned effects into account. The interferer (Comm) is a four-transmitter node. The discrepancy between the measured and simulated SNIR_{out} after interference cancellation was 1 dB, which can be attributed to hardware imperfections and complex multipath propagation, which were not put into the simulation. This validates that the interference cancellation algorithm is employable with real signals and is able to separate the Comm signals adequately to significantly improve the SNIR_{out} for a reliable radar estimation.
- Outdoor measurement verifications featuring static targets verified the unambiguous angular regions (antenna array dependent) and showed the effect of target estimation ambiguity when this region is breached.
- Another outdoor measurement captured a van moving at around 30 km/h away from the Radar. The effect of interference from a one-transmitter Comm node and the interference canceled results were also presented to wrap up the validation of the interference cancellation algorithm proposed in this work.

With all the theoretical findings, algorithm, strategy and measurement verifications presented in this work, the goal toward a dual-functioning radar and communication system is now one step closer to being realized. While more researches can be done to further improve the radar estimation through strategic placement of multiple RadCom nodes or post-processing of the radar processed matrices, it is now clear that a real-time MIMO RadCom is definitely implementable with the current technology that is available. If enough financial and cooperation support from various bodies i.e. universities, industries and government can be garnered, it can be anticipated that within five years we will be able to see MIMO RadComs used in the automotive and security sectors as predicted by Wiesbeck et al. in [WSY+15].

Bibliography

[AMGC02] M.S. Arulampalam, S. Maskell, N. Gordon, and T. Clapp. A Tutorial on Particle Filters for Online Nonlinear/Non-Gaussian Bayesian Tracking. *IEEE Trans. on Signal Proc.*, 50(2):174–188, Feb 2002.

[Bal05] Constantine A. Balanis. *Antenna Theory: Analysis and Design.* Wiley-Interscience, 2005.

[Bas16] Sanjoy Basak. *System Setup and Calibration using USRPs for Parallel Transmit-Receive MIMO Radar Operation.* Master thesis, Institut für Hochstfrequenztechnik und Elektronik of Karlsruhe Institute of Technology, Karlsruhe, Germany, 2016.

[BF92] James R. Bunch and Ricardo D. Fierro. A constant-false-alarm-rate algorithm. *Linear Algebra and its Applications*, 172:231 – 241, 1992.

[BFD+09] D.W. Bliss, K.W. Forsythe, S.K. Davis, G.S. Fawcett, D.J. Rabideau, L.L. Horowitz, and S. Kraut. GMTI MIMO radar. In *International Waveform Diversity and Design Conference*, pages 118–122, Feb 2009.

[BP99] S. Blackman and R. Popoli. *Design and Analysis of Modern Tracking Systems.* Artech House, 1999.

[Bra14] Klaus Martin Braun. *OFDM Radar Algorithms in Mobile Communication Networks.* PhD thesis, Communications Engineering Lab of Karlsruhe Institute of Technology, Karlsruhe, Germany, 2014.

[Bro13] E. Brookner. MIMO Radar Demystified and Where It makes Sense to use. In *IEEE International Symposium on Phased Array Systems Technology*, pages 399–407, Oct 2013.

[Bro15] E. Brookner. MIMO Radars and Their Conventional Equivalents. In *IEEE Radar Conference (RadarCon)*, pages 0918–0924, May 2015.

[BSNJ09] M. Braun, C. Sturm, A. Niethammer, and F.K. Jondral. Parametrization of Joint OFDM-based Radar and Communication Systems for Vehicular Applications. In *IEEE 20th International Symposium on Personal, Indoor and Mobile Radio Communications; 2009*, pages 3020 –3024, Sep. 2009.

[BTJ13] Martin Braun, Ralph Tanbourgi, and Friedrich K. Jondral. Co-channel Interference limitations of OFDM Communication-Radar Networks. *EURASIP Journal on Wireless Communications and Networking*, 2013(1):1–16, 2013.

[Cao04] Tri-Tan Van Cao. A CFAR algorithm for radar detection under severe interference,. In *Proceedings of the 2004 Intelligent Sensors, Sensor Networks and Information Processing Conference, 2004*, 2004.

[car20] *Mutiple antennas a requirement for new cars*, Accessed: 2016-03-20. `http://www.edn.com/electronics-blogs/automotive-innovation/4421619/Multiple-antennas-a-requirement-for-new-cars`.

[Cav91] J. K. Cavers. An Analysis of Pilot Symbol Assisted Modulation for Rayleigh Fading Channels [mobile radio]. *IEEE Transactions on Vehicular Technology*, 40(4):686–693, Nov 1991.

[CMB08] C.Williams, S. McLaughlin, and M. A. Beach. Robust OFDM Timing Synchronisation in Multipath Channels. *EURASIP Journal on Wireless Communications and Networking*, 2008(675048), April. 2008.

[DH09] F. Daum and J. Huang. MIMO Radar: Snake oil or Good Idea? In *International Waveform Diversity and Design Conference, 2009*, pages 113–117, Feb 2009.

[Doh03] Mischa Dohler. *Virtual Antenna Arrays*. PhD thesis, Department of Electrical and Electronic Engineering, King's College London, University of London, Nov. 2003.

[dou30] *Double-ridged Horn Antennas from ETS-Lindgren*, Accessed: 2016-03-30. http://www.ets-lindgren.com/pdf/3115.pdf.

[dri10] *DriveC2X Project*, Accessed: 2016-07-10. http://www.drive-c2x.eu/project.

[ett20a] *ETTUS SBX daughterboard performance data*, Accessed: 2016-03-20. http://files.ettus.com/performance_data/sbx/SBX-without-UHD-corrections.pdf.

[ett20b] *USRP Hardware Driver and USRP Manual: General Application Notes*, Accessed: 2016-03-20. http://files.ettus.com/manual/page_general.html.

[FCL12] P. Falcone, F. Colone, and P. Lombardo. Localization of Moving Targets with a Passive Radar System Based on WiFi Transmissions. In *IET International Conference on Radar Systems (Radar 2012)*, pages 1–6, Oct 2012.

[FHB+04] E. Fishler, A. Haimovich, R. Blum, D. Chizhik, L. Cimini, and R. Valenzuela. MIMO Radar: An Idea Whose Time has Come. In *Proceedings of the IEEE Radar Conference*, pages 71–78, April 2004.

[Fin86] H. M. Finn. A CFAR Design for a Window Spanning Two Clutter Fields. *IEEE Transactions on Aerospace and Electronic Systems*, AES-22:155–169, March 1986.

[FNvG06] G.E.A. Franken, H. Nikookar, and P. van Genderen. Doppler Tolerance of OFDM-coded Radar Signals. In *3rd European Radar Conference (EURAD)*, pages 108 –111, Sept. 2006.

[fre20] *Freescale Semiconductor Application Note: Channel Estimation in OFDM Systems by Shen and Martinez*, Accessed: 2016-03-20. http://www.nxp.com/files/dsp/doc/app_note/AN3059.pdf.

[GG03] E. F. Greneker and J. L. Geisheimer. The Use of Passive Radar for Mapping Lightning Channels in a Thunderstorm. In *Proceedings of the 2003 IEEE Radar Conference, 2003*, pages 28–33, May 2003.

[GSKS09] D. Garmatyuk, J. Schuerger, K. Kauffman, and S. Spalding. Wideband OFDM System for Radar and Communications. In *IEEE Radar Conference.*, pages 1 –6, May 2009.

[GW10] H. Griffiths and N. Willis. Klein Heidelberg - The First Modern Bistatic Radar System. *IEEE Transactions on Aerospace and Electronic Systems*, 46(4):1571–1588, Oct 2010.

[Har78] F. J. Harris. On the Use of Windows for Harmonic Analysis with the Discrete Fourier Transform. *Proceedings of the IEEE*, 66(1):51–83, Jan 1978.

[HMS+13] Marlene Harter, Tobias Mahler, Tom Schipper, Andreas Ziroff, and Thomas Zwick. 2-D Antenna Array Geometries for MIMO Radar Imaging by Digital Beamforming. In *European Microwave Conference (EuMC); 2013*, pages 1695–1698, 2013.

[HSL09] Hao He, P. Stoica, and Jian Li. Unimodular Sequence Sets with Good Correlations for MIMO Radar. In *IEEE Radar Conference; 2009*, pages 1–6, May 2009.

[HZZ11] M. Harter, A. Ziroff, and T. Zwick. Three-dimensional radar imaging by digital beamforming. In *European Radar Conference (EuRAD); 2011*, pages 17–20, Oct 2011.

[Jun04] Wang Jun. A Parametric Scattering Centers Model with Frequency Dependence for Radar Target Recognition. In *Asia-Pacific Radio Science Conference, 2004. Proceedings. 2004*, pages 98–100, Aug 2004.

[khr10] *Klein-Heiderberg a Quite Mysterious Bi-Static Radar System*, Accessed: 2016-04-10. `http://www.cdvandt.org/k-h.htm`.

[Kno12] E.F. Knott. *Radar Cross Section Measurements*. Springer US, 2012.

[Lev00] N. Levanon. Multifrequency Complementary Phase-Coded Radar Signal. *IEE Proceedings - Radar, Sonar and Navigation*, 147(6):276 –284, Dec. 2000.

[LS08] Jian Li and Petre Stoica. *MIMO Radar — Diversity Means Superiority*, pages 1–64. John Wiley & Sons, Inc., 2008.

[LSX06] J. Li, P. Stoica, and Y. Xie. On Probing Signal Design for MIMO Radar. In *Fortieth Asilomar Conference on Signals, Systems and Computers; 2006*, pages 31–35, Oct 2006.

[MM99] M. Morelli and U. Mengali. An Improved Frequency Offset Estimator for OFDM Applications. In *Communication Theory Mini-Conference*, pages 106–109, Jun 1999.

[mos20] *More Safety for All by Radar Interference Mitigation (MOSARIM), 7th Frame EU Project, FP7/2007-2013*, Accessed: 2016-03-20. `http://www.mosarim.eu`.

[MZB00] H. Minn, M. Zeng, and V. K. Bhargava. On Timing Offset Estimation for OFDM Systems. *IEEE Communications Letters*, 4(7):242–244, July 2000.

[Ngu12] Thanh Thuy Nguyen. *Design and Analysis of a Superresolution Algorithm and Signal Separation Technique for an OFDM-based MIMO Radar*. Master thesis, Institut für Hochstfrequenztechnik und Elektronik of Karlsruhe Institute of Technology, Karlsruhe, Germany, 2012.

[NSZ16] B. Nuss, Y.L. Sit, and T. Zwick. 3D Radar Image Fusion using OFDM-based MIMO Radar. In *The 10th German Microwave Conference (GeMiC), 2016*, March 2016.

[OA07] M. K. Ozdemir and H. Arslan. Channel Estimation for Wireless OFDM Systems. *IEEE Communications Surveys Tutorials*, 9(2):18–48, Second 2007.

[Orz16] Mateusz Orzol. *System Setup and Calibration using Universal Software Radio Peripherals with Antenna Arrays for Multiple-Input Multiple-Output Radar Direction-of-Arrival Estimation*. Bachelor thesis, Institut für Hochstfrequenztechnik und Elektronik of Karlsruhe Institute of Technology, Karlsruhe, Germany, 2016.

[PFS15] C. Pfeffer, R. Feger, and A. Stelzer. A Stepped-Carrier 77-GHz OFDM MIMO Radar System with 4 GHz Bandwidth. In *European Radar Conference (EuRAD); 2015*, pages 97–100, Sept 2015.

[PIK14] Idnin Pasya, Naohiko Iwakiri, and Takehiko Kobayashi. Joint Direction-of-Departure and Direction-of-Arrival Estimation in a UWB MIMO Radar Detecting Targets with Fluctuating Radar Cross Sections. *International Journal of Antennas and Propagation*, 2014(847815):242–244, 2014.

[PRK86] A. Paulraj, R. Roy, and T. Kailath. A Subspace Rotation Approach to Signal Parameter Estimation. *Proceedings of the IEEE*, 74(7):1044 – 1046, Jul. 1986.

[RFZ09] L. Reichardt, T. Fugen, and T. Zwick. Influence of Antennas Placement on Car to Car Communications Channel. In *3rd European Conference on Antennas and Propagation; 2009*, pages 630–634, March 2009.

[Rob47] S. D. Robertson. Targets for Microwave Radar Navigation. *The Bell System Technical Journal*, 26(4):852–869, Oct 1947.

[RPK86] R. Roy, A. Paulraj, and T. Kailath. ESPRIT–A Subspace Rotation Approach to Estimation of Parameters of Cisoids in Noise. *IEEE Transactions on Acoustics, Speech and Signal Processing*, 34(5):1340–1342, Oct 1986.

[RSGZ12] L. Reichardt, C. Sturm, F. Grünhaupt, and T. Zwick. Demonstrating the Use of the IEEE 802.11P Car-to-Car Communication Standard for Automotive Radar. In *6th European Conference on Antennas and Propagation (EUCAP); 2012*, pages 1576–1580, March 2012.

[SBZW10] Christian Sturm, Martin Braun, Thomas Zwick, and Werner Wiesbeck. A System Demonstrator for the Performance Evaluation of a 24 GHz ISM Band Radar operating with OFDM Waveforms. In *IEEE Antennas and Propagation Society International Symposium (APSURSI); 2010*, pages 1 –4, Jul. 2010.

[SC97] T.M. Schmidl and D.C. Cox. Robust Frequency and Timing Synchronization for OFDM. *IEEE Transactions on Communications*, 45(12):1613 –1621, Dec. 1997.

[Sch86] R. Schmidt. Multiple Emitter Location and Signal Parameter Estimation. *IEEE Transactions on Antennas and Propagation*, 34(3):276 – 280, Mar. 1986.

[Sch07] Karin Schuler. *Intelligente Antennensysteme für Kraftfahrzeug-Nahbereichs-Radar-Sensorik.* PhD thesis, Institut für Hochstfrequenztechnik und Elektronik der Universität Karlsruhe (TH), 2007.

[sem20] *The Physiology of Semi-Autonomy and Test Results*, Accessed: 2016-05-20. `http://www.caranddriver.com/features/semi-autonomous-cars-compared-tesla-vs-bmw-mercedes-and-infiniti-feature-the-physiology-of-semi-autonomy-and-test-results-page-6`.

[Sie58] K. M. Siegel. Bistatic Radars and Forward Scattering. In *Proc. National Conference on Aerospace and Electronics, Dayton, Ohio*, pages 286–290, 1958.

[SN09] S. Sen and A. Nehorai. OFDM MIMO Radar for Low-Grazing Angle Tracking. In *Conference Record of the Forty-Third Asilomar Conference on Signals, Systems and Computers; 2009*, pages 125–129, Nov 2009.

[SNB+16] Y. L. Sit, B. Nuss, S. Basak, M. Orzol, and T. Zwick. Demonstration of Interference Cancellation in a Multiple-User Access OFDM MIMO Radar-Communication Network using USRPs. In *IEEE International Conference on Microwave for Intelligent Mobility (ICMIM); 2016*, May 2016.

[SNSZ13] Y. L. Sit, T. T. Nguyen, C. Sturm, and T. Zwick. 2D Radar Imaging with Velocity Estimation using a MIMO OFDM-Based Radar for Automotive Applications. In *European Radar Conference (EuRAD); 2013*, pages 145–148, Oct 2013.

[SPZW09] C. Sturm, E. Pancera, T. Zwick, and W. Wiesbeck. A Novel Approach to OFDM Radar Processing. In *IEEE Radar Conference; 2009*, pages 1–4, May 2009.

[SS91] B.D. Steinberg and H.M. Subbaram. *Microwave Imaging Techniques*. Wiley Series in Remote Sensing and Image Processing. Wiley, 1991.

[SSBZ13] C. Sturm, Y. L. Sit, M. Braun, and T. Zwick. Spectrally Interleaved Multi-Carrier Signals for Radar Network Applications and Multi-Input Multi-Output Radar. *IET Radar, Sonar Navigation*, 7(3):261–269, 2013.

[SSWZ15] Y. L. Sit, B. Sobhani, W. Wiesbeck, and T. Zwick. Simple OFDM-Based MIMO Radar for Real-Time Short-Range Area Surveillance. In *European Radar Conference (EuRAD); 2015*, pages 121–124, Sept 2015.

[SSZ11] Y. L. Sit, C. Sturm, and T. Zwick. Interference Cancellation for Dynamic Range Improvement in an OFDM Joint Radar and Communication System. In *European Radar Conference (EuRAD), 2011*, pages 333–336, Oct 2011.

[SSZ12] Y. L. Sit, C. Sturm, and T. Zwick. One-Stage Selective Interference Cancellation for the OFDM Joint Radar-Communication System. In *The 7th German Microwave Conference (GeMiC), 2012*, pages 1–4, March 2012.

[Stu12] Christian Andreas Sturm. *Gemeinsame Realisierung von Radar-Sensorik und Funkkommunikation mit OFDM-Signalen*. PhD thesis, Institut für Hochstfrequenztechnik und Elektronik der Karlsruhe Institut für Technologie, Karlsruhe, Germany, 2012.

[SW11] C. Sturm and W. Wiesbeck. Waveform Design and Signal Processing Aspects for Fusion of Wireless Communications and Radar Sensing. *Proceedings of the IEEE*, 99(7):1236–1259, July 2011.

[SZ14] Y. L. Sit and T. Zwick. MIMO OFDM Radar with Communication and Interference Cancellation Features. In *IEEE Radar Conference; 2014*, pages 0265–0268, May 2014.

[SZWB10] C. Sturm, T. Zwick, W. Wiesbeck, and M. Braun. Performance Verification of Symbol-based OFDM Radar Processing. In *IEEE Radar Conference; 2010*, pages 60 –63, May. 2010.

[tes20a] *Tesla Motors: Super charger*, Accessed: 2016-03-20. `https://www.teslamotors.com/supercharger`.

[tes20b] *Tesla Motors: Your Autopilot has Arrived*, Accessed: 2016-03-20. `https://www.teslamotors.com/blog/your-autopilot-has-arrived`.

[vdBSB97] J. J. van de Beek, M. Sandell, and P. O. Borjesson. ML Estimation of Time and Frequency Offset in OFDM Systems. *IEEE Transactions on Signal Processing*, 45(7):1800–1805, Jul 1997.

[Wei10] F. Weinmann. Frequency Dependent RCS of a Generic Airborne Target. In *URSI International Symposium on Electromagnetic Theory (EMTS), 2010*, pages 977–980, Aug 2010.

[who20a] *World Health Oraganization (WHO): Global status report on road safety 20152*, Accessed: 2016-03-20. `http://www.who.int/violence_injury_prevention/road_safety_status/2015/GSRRS2015_Summary_EN_final2.pdf`.

[who20b] *World Health Oraganization (WHO): The 10 leading causes of death in the world, 2000 and 2012*, Accessed: 2016-03-20. `http://www.who.int/mediacentre/factsheets/fs310/en/`.

[WKG10] X. H. Wu, A. A. Kishk, and A. W. Glisson. MIMO-OFDM Radar for Direction Estimation. *IET Radar, Sonar Navigation*, 4(1):28–36, February 2010.

[WMR10] Q. Wang, C. Mehlfuhrer, and M. Rupp. Carrier Frequency Synchronization in the Downlink of 3GPP LTE. In *IEEE 21st International Symposium on Personal Indoor and Mobile Radio Communications (PIMRC); 2010*, pages 939–944, Sept 2010.

[WRF+94] J. Wurman, M. Randall, C. L. Frush, E. Loew, and C. L. Holloway. Design of a Bistatic Dual-Doppler Radar for Retrieving Vector Winds using one Transmitter and a Remote Low-Gain Passive Receiver. *Proceedings of the IEEE*, 82(12):1861–1872, Dec 1994.

[WSY+15] W. Wiesbeck, L. Sit, M. Younis, T. Rommel, G. Krieger, and A. Moreira. Radar 2020: The Future of Radar Systems. In *IEEE International Geoscience and Remote Sensing Symposium (IGARSS)*, pages 188–191, July 2015.

[ZS13] T. Zwick and C. Sturm. Method for Separating Transmitted Signals in a Radar System, December 26 2013. US Patent App. 13/361,038.